POSTPRINCIPIA
Gravitation for Physicists and Astronomers

Peter Rastall was born in Lincolnshire, twenty-five kilometers from the birthplace of Isaac Newton. He was educated at Lincoln School and the University of Manchester, where he studied theoretical physics. He worked for two years in the Aerodynamics Department of Rolls-Royce, and then emigrated to Canada. Apart from years spent in Australia, Texas, and Colorado, he has since lived in Vancouver, where he is a professor in the Physics Department of the University of British Columbia.

POST PRIN CIPIA

GRAVITATION for PHYSICISTS and ASTRONOMERS

PETER RASTALL
University of British Columbia

World Scientific
Singapore • New Jersey • London • Hong Kong

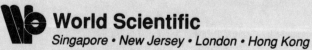

Published by

World Scientific Publishing Co. Pte. Ltd.

P O Box 128, Farrer Road, Singapore 9128

USA office: Suite 1B, 1060 Main Street, River Edge, NJ 07661

UK office: 73 Lynton Mead, Totteridge, London N20 8DH

POSTPRINCIPIA
Gravitation for Physicists and Astronomers

ISBN 981-02-0778-6

Printed in Singapore by General Printing Services Pte. Ltd.

Foreword

Newtonian gravity is a dead subject, in the opinion of most physicists. The best that a book can do is embalm it prettily for the sake of necrophilic astronomers. The introductory paragraphs, suitably ponderous, should go something like this.

> *The Newtonian theory of gravity is one of the great triumphs of science. It gives a simple prescription for calculating the motions of heavenly bodies that is in almost complete accord with observation. The agreement is not quite perfect – the motion of Mercury is slightly anomalous – but for more than two hundred years the theory was never seriously challenged.*
>
> *With the discovery of special relativity at the beginning of this century, it became obvious that the Newtonian theory is, at best, incomplete. Just as electrostatics was incorporated into the relativistically acceptable electrodynamics of Maxwell, so must the Newtonian theory of gravitostatics be incorporated into some relativistic theory of gravity.*
>
> *Although the generalization of electrostatics was easy, at least in retrospect, that of gravitostatics is not. It requires a fundamental change in our ideas about spacetime, and quite sophisticated mathematics, which are unfortunately beyond the scope of the present book. Instead, etc.*

No reasonable person would disagree with the the first two paragraphs. In the third, our imaginary author throws in his hand, and says that he cannot really explain gravity to us until we are expert in differential geometry. He then, no doubt, devotes the rest of his preface and his book to the delights of the Kepler problem and perturbation theory.[1,2]

---

1 Superscript numbers indicate notes, which are collected at the end of each chapter. They are seldom of such importance that one should page forward to find them. References are given in brackets, with the (first) author's name and the last two digits of the year of publication, e.g. [Newton 86]. They are listed near the end of the book.

2 There are real authors who can resurrect even these dry bones – for example, [Taff 85].

Before we decide to follow him, let us digress to a field where a similar difficulty is avoided.

A proper, *mathematical* treatment of quantum mechanics requires a knowledge of Hilbert space, which is usually not inflicted on physicists until they reach graduate school. This does not mean that undergraduate quantum mechanics is reduced to the photoelectric effect and the Bohr atom. It is possible to acquire a good understanding of the subject, and to solve a wide variety of problems, with quite elementary mathematics – and there are many books to show one how.

We may ask why nothing comparable has occurred for gravity. It is partly because, until recently, there have been very few experimental results that could not be explained by the elementary theory. Perhaps too, there was reluctance on the part of gravitational theorists to admit that their hard-won geometrical methods are not always essential. A few authors tried to fill the gap by producing simplified versions of the geometrical theory suitable for theoretical graduate students, or even for mathematical undergraduates [Weinberg 72], [Schutz 85]. But there is no textbook for unmathematical undergraduates, or for experimentalists who are impatient of geometrical subtleties. I hope that this book will be suitable for both groups, and also for general readers who wish to extend their knowledge beyond Newtonian bounds. For the mathematical relativist, it may be amusing to peer into the abyss of physical thinking.

The first half of the book generalizes the Newtonian theory, in two stages. In Chapters I–IV, gravitostatics is developed – the theory of static or quasistatic gravitational fields. This already enables one to calculate all the "classical tests" of relativistic gravity. The treatment is suitable for physics or astronomy students in their second or third undergraduate year. No previous knowledge of special relativity is necessary, although a small amount might be helpful. Appendix A is a heuristic introduction, or reintroduction, to relativity for those who feel that they are intolerably ignorant.

The second stage of generalization, Chapters V and VI, is to the postnewtonian theory, which is sufficiently precise for most calculations in astrophysics and astronomy. The physics is no harder than in the preceding chapters, but the equations are sometimes longer. The theory is applied to fluid mechanics in Chapter VII and to the motions of systems of bodies in Chapter VIII. The material is suitable for senior undergraduates or beginning graduate students. They will find that the requisite relativity is summarized in Appendix B, and the potential theory in Appendix C.

The main, subversive aim of the book is to show how the postnewtonian approximation can be derived, with all the correct parameters, by a natural generalization of the Newtonian theory. The derivation is important because almost all the empirical evidence for exact, relativistic theories of gravity is at the postnewtonian level. If the postnewtonian approximation can be derived from the Newtonian theory, it follows that this evidence cannot be adduced in favour of any particular relativistic theory. The best that one can say at present of any such theory is that it is as good as the Newtonian – and perhaps that it accounts for the motion of binary pulsars.

Our nonsubversive aim is to assist those who struggle with postnewtonian calculations. The treatises on the subject are rather heavy reading because they deal, as they should, with the general theory in all its complexity [Brumberg 72], [Will 81], [Soffel 89]. Our treatment is simpler, and sufficient for many purposes. The notation is very similar to Will's and Soffel's, and this book can serve as an easy introduction to theirs.

One might hope that the path of plausible generalization which leads to the postnewtonian theory would extend further and allow the calculation of gravitational radiation and the motion of very massive bodies. Chapter IX shows that such a generalization is not possible: the postnewtonian theory is as far as one can go without substantial, new hypotheses. We establish this by describing two plausible, but quite different, exact theories of gravity that reduce to the postnewtonian theory in the appropriate limit. (There is, presumably, an infinite number of such theories, but two are sufficient for our purpose!)

The lack of a unique, exact generalization does not imply that the postnewtonian theory plays no part in the calculation of gravitational radiation, etc. We show that it has an important role, and is an essential intermediary between exact theories and observation. But a proper treatment of gravitational radiation requires another, and more mathematical book.

This text is different from most and may appear eccentric, but is in fact not so. It is the main stream that has been diverted from its proper bed. The physics that we advocate is sober and conservative. We drive a well established theory as far as possible, until it breaks down or runs out of fuel – and only then do we look for a new one. The alternative, followed by almost all textbooks on gravity, is to give up at the first sign of trouble. They abandon the Newtonian theory for no good reason: there are a few small things wrong, but they do not try to fix them. They invite the reader to spend half a year in learning differential geometry, and promise that he will then be given a new theory that is more satisfactory. Most physicists are, understandably, unwilling to take this path. If they do, they discover that the new theory is hard to apply in practical calculations, except in its approximate, linear or postnewtonian, forms. From a logical and pedagogical point of view, the course that we follow is undoubtedly the best. One first understands the postnewtonian theory and the physical basis of gravity, and then ascends to more abstract realms.

Physicists today have far less confidence in the validity of the extant relativistic theories of gravity than they had a decade or two ago. Cosmological perplexities and the inflationary extravaganzas invoked to explain them, the "missing mass" that obstinately refuses to be found, and the competition between different theories, all foster scepticism. One might conclude that relativistic theories of gravity should be left to the specialists, as being too speculative for the standard graduate curriculum. However, there is a place for speculation that is not too earnest – for tongue-in-cheek fantasy that may sometimes lead to more serious probing. This book provides a solid basis for such philandering, as well as for dour calculation.

Acknowledgments

The first part of the book was written at the University of Western Australia. I thank the Department of Physics for the opportunity to work undisturbed in that pleasant place. The typescript was produced with the aid of the mathematical word processor ChiWriter. It was turned into elegant TeX by Janet Clark, who is responsible for all the algebraic errors.

A note on style

The choice of a proper style for scientific writing is not easy. Most textbooks are a uniform grey. The reader is challenged to stay awake long enough to find the treasures that may be hidden below the muddy surface. Something more lively is called for – but it must not distract from the content. Although there are many detailed arguments to go through that cannot be made exciting line by line, one should at least try to change the pace, or tell occasional jokes, or make outrageous errors (to be fair, the last method is widely used – even in the dullest texts). I have attempted to steer between the dry and the outlandish. It is a narrow way beset by failures of taste, for which I ask the reader's pardon.

Contents

I Newtonian Gravity

1. Introduction

The ideas of physics come muddy footed from an impure world. Although in time they may be cleaned and dressed mathematically, they keep their power only if we remember their origins. To understand physics is not just to master abstract mathematical relationships, but to comprehend something of the process of abstraction.

It would be logical to begin at the beginning. We could dream of creating a clean, mathematical structure to which no idea would be admitted unless it wiped its feet. With great effort we might build such an edifice – and it would be impregnable to all but professional logicians. We shall therefore be less ambitious. Like most books for physicists, this one begins in the middle: not at the most abstract level, but in a realm where we are already accustomed to think.

We first briefly expound the Newtonian theory in its familiar form. Then, to extend the theory, we travel in two directions – forwards to greater generality and precision, and also back along the path we thought we knew, to reassess the relationship between theory and observation. We are adrift on a raft of inherited notions in a sea of confusion. There is no shore where we might build a new ship, free from misconceptions, or even pick up pebbles. We can only tighten the lashings, replace rotten logs, and throw overboard objects of doubtful value. Beginning with these extravagant metaphors.

2. Review of the Newtonian theory

The Newtonian theory – by which we mean Newton's original theory together with eighteenth and nineteenth century additions – is the standard, prerelativistic theory of gravity. The reader should already know its main features, which are summarized below for reference and quick review.

Space is Euclidean and 3 dimensional

There is a deliberate ambiguity here. Euclidean space is a mathematical object. It is a set of elements (called *spatial points* or simply *points*) that satisfy a number of axioms. Points are *not* empirical (i.e. observable or measurable) objects, and if one speaks of the distance between points, for example, one is talking mathematics.

Space, in its other meaning, is physical space, and is inhabited by material objects. We may be able to measure the distance between two small, material objects, by using a measuring rod, or radar, or some other means. If we can also make a correspondence between the material objects and a pair of points, then we can consider the relationship between the (empirical, or measured) distance between the objects and the (mathematical, or Euclidean) distance between the points. The simplest case, and the only one considered by the Newtonian theory, is when one distance is approximately equal to the other.

To say that space is *3 dimensional* means that any spatial point P can be unambiguously labelled by an ordered triple of real numbers $\boldsymbol{x}(P) = (x^1(P), x^2(P), x^3(P))$. We say that $\boldsymbol{x}$ is a *chart* or *coordinate system*, and that $\boldsymbol{x}(P)$ (or the real numbers $x^m(P)$, $m \in \{1, 2, 3\}$) are the *coordinates* of the spatial point P in $\boldsymbol{x}$.[1]

There are charts, called *general cartesian charts*, in which the distance $\overline{PQ}$ between any pair of spatial points P and Q is given by

$$\overline{PQ} = \sigma \mid \boldsymbol{x}(P) - \boldsymbol{x}(Q) \mid = \sigma \Big[\sum_m [x^m(P) - x^m(Q)]^2 \Big]^{1/2}, \tag{2.1}$$

where m is summed over $\{1, 2, 3\}$, the non-negative value of the square root is taken, and σ is a strictly positive constant (the same for all P and Q). In the particular case when $\sigma = 1$, the chart is a *cartesian chart*. The coordinates of a (general) cartesian chart are called *(general) cartesian coordinates*. They are assigned the dimension of length, because the Euclidean distance $\overline{PQ}$ may correspond to an empirically measurable length – a distance between small material objects – in the manner we have described. In what follows, all charts are cartesian unless stated otherwise. General cartesian charts with $\sigma \neq 1$ will occasionally appear in the next chapter.

Existence of a time coordinate

Again there is an ambiguity, because *time* has both a mathematical and an empirical meaning. Mathematically speaking, the time coordinate is a real variable that may be used to label mathematical objects; its deeper significance will not emerge until later. The time coordinate is arbitrary to the extent of an affine transformation $t \mapsto at + b$, where a and b are constants, $a \neq 0$.

There is a rich time-vocabulary that we shall draw on quite freely. We may speak of the *instant* t, instead of the *time* t. A *time interval* is the set of values of the time variable that are contained in an interval of $\mathbb{R}$. A *point-event*[2] is an ordered pair (P, t), and two point-events (P, t) and (P', t') are *simultaneous* iff $t = t'$ (recall that *iff* means *if and only if*: two-way implication).

In the empirical sense, time is a quantity measured by an ideal clock (one says *ideal* to avoid lengthy discussions of the idiosyncrasies of real clocks). In some situations, empirical time may be related to the mathematical time variable – again the only relationship considered by the Newtonian theory is approximate equality (apart from a possible affine transformation). For example, one may be able to associate a clock with a small material object, and hence with a spatial point P. Something happening close to the object at a measured time t (which may be called a *physical point-event*) can then be associated with the point-event (P, t).

The Newtonian theory assumes that two clocks can always be synchronized (i.e. set to read the same time), and that once synchronized they stay synchronized. The rectification of this mistaken notion is a concern of the theory of relativity. In the next chapter, we shall criticize it from another point of view.

Particles and paths

A small material body is a *particle* (or, more explicitly, an *empirical particle*) if its dimensions and structure are of no importance in a specified situation. This means that a given material body may sometimes be regarded as a particle and sometimes not. For example, two gravitationally interacting stars may be regarded as particles when their distance apart is large in comparison with their diameters, but not when they are close together.

If an empirical particle is associated with the point P at the time t (i.e. at the measured time that corresponds to the value t of the mathematical time coordinate), we can say that it is associated with the point-event (P, t). An empirical particle may be associated with different points at different times. The points lie on a curve, which we call the *path* of the particle.

More precisely, we define the path of a mathematical particle to be a curve in 3 space. It is a map $\boldsymbol{z}$ from the real numbers $\mathbb{R}$ into the set of spatial points $\mathcal{S}$. We write $\boldsymbol{z} : \mathbb{R} \to \mathcal{S}$, $\boldsymbol{z}(t) = P$, and say that P is the spatial point corresponding to the time t. A mathematical particle is then identified with $\boldsymbol{z}$, together with a set of parameters (mass, charge, spin angular momentum, etc.) that determine its properties. In the simplest case, one might denote the particle as $(\boldsymbol{z}, m)$, where m is a scalar quantity – the *mass* of the particle.

In the by now familiar manner, we may speak of particles without specifying whether they are mathematical or empirical – with the usual warning that caution is necessary. Too naive an identification may lead to nonsensical assertions.

We note that there is an alternative definition of a curve in terms of the coordinates of points, rather than the points themselves. A curve is then a map

$Z : \mathbb{R} \to \mathbb{R}^3$, $Z(t) = (Z^1(t), Z^2(t), Z^3(t))$, where $Z^m(t)$ is interpreted to be the m coordinate of a point P in a cartesian chart x (that is, $Z(t) = x(P)$, where P depends on t). The previous definition has the advantage of not depending on any particular choice of cartesian chart; the new one allows us to define continuity and differentiability of the curve very easily, in terms of the continuity and differentiability of the functions Z^m. For example, we say that the curve is C^0 if the Z^m are continuous, for $m \in \{1, 2, 3\}$; it is C^1 if the Z^m have continuous first derivatives; it is C^r, for any positive integer r, if the Z^m have continuous r^{th} derivatives. We shall usually assume that the paths of particles are C^2 curves. In either of the definitions of a curve, one can replace the domain $\mathbb{R}$ by an open interval $\mathbb{I} \subset \mathbb{R}$ (and write $z : \mathbb{I} \to S$, for example).

The use of coordinates rather than points is often convenient in other contexts. If x is a cartesian chart, we write $x = (x^1, x^2, x^3)$, where the x^m, with $m \in \{1, 2, 3\}$, are the *coordinate functions* of x; the coordinates of the spatial point P in x are then $x^m(P)$. However, we shall often follow the practice of physicists and *identify* a spatial point with a triple of real numbers. That is, we speak of a point (x^1, x^2, x^3), where the x^m are real numbers and not maps from the set S of spatial points into $\mathbb{R}$.

Forces

We have sketched a picture of the world as a swarm of particles following their paths. In the jargon of the trade, it is a *kinematical* rather than a *dynamical* description. (The difference is like that between story – "This happened and that happened", and plot – "This happened *because* that happened".) We must now ask why certain paths are followed in preference to others. We have to consider the interactions between particles, and examine the Newtonian idea of force.

The idea of force arises from everyday experience, and was incorporated into physics very early (statics, etc.). Newton's achievement was to use the idea to give a general prescription for the solution of dynamical problems. We shall present only the barest summary; for more details, the historians must be consulted [Westfall 71].

Contemplating the universe, we see a web of interactions of incalculable complexity. How can we understand anything until we understand everything? Newton's answer, although not in these words, is that one must first divide the universe into two parts: the *system* and the *environment*. (The division is conceptual only – no material barriers need be erected.) One then assumes – the crucial step – that the effect of the environment on the system is equivalent to a set of forces. In other words, the environment can be *replaced* by a set of forces so far as the system is concerned.

We are taking the point of view that the system consists of a set of particles, and the forces are then vector fields defined on the paths. The motion of the system is found by solving differential equations that involve the forces, but are otherwise independent of the environment.

It is not, of course, the whole story. We still have to determine the forces, and the equations of motion (the differential equations that relate the forces to the variables of the system). These are not trivial questions, but they are fairly well defined. What Newton has given us is a uniform way of approaching dynamical problems – of breaking them into smaller and more tractable ones.

The reader may feel that what has been said is obvious to the point of triviality. We are so accustomed to attack problems in this manner that we have difficulty in imagining any other possibility; we find it hard to believe that Newton had to *invent* the method. A truly great discovery is perhaps one that makes previous ideas unthinkable.... But we have spent enough time on the pseudophilosophical plateaux. Let us descend to more fertile lands.

Empirical force, as a measurable quantity, is slipperier than distance or time. Even in such a simple example as the force measured by a spring balance, one can argue that what is really measured is a distance (the position of the pointer).

Mathematically speaking, the force on a particle at an instant t is an element of a 3 dimensional vector space. We shall usually assume that there is a differentiable vector field $\boldsymbol{F}$ whose value $\boldsymbol{F}(t)$ is the force on the particle at the instant t, for all t in some interval. In the sloppy manner of physicists, we may regard force as a vector or a vector field, as convenient.

Equations of motion

We have described two ways of representing a curve. The second way – as a map $\boldsymbol{Z} : \mathbb{I} \to \mathbb{R}^3$, $\boldsymbol{Z}(t) = (Z^1(t), Z^2(t), Z^3(t))$, where $Z^m(t)$ is the m coordinate of a point in a cartesian chart x - will be called a *vector* representation. We are here treating vectors in a very primitive way, as ordered triples of real numbers. Later, when we consider the relationship between different charts, we may have to be more sophisticated.

In terms of a vector representation, we define the *velocity* $\boldsymbol{V}$ and the *acceleration* $\boldsymbol{A}$ of a particle by $\boldsymbol{V} = D\boldsymbol{Z}$ and $\boldsymbol{A} = D\boldsymbol{V}$, where D denotes the derivative. We write the components of $\boldsymbol{V}$ as $V^r = DZ^r$, etc. The *momentum* of a particle with velocity $\boldsymbol{V}$ and constant mass m is defined to be $m\boldsymbol{V}$. Although non-constant masses are permissible in the Newtonian theory, we do not consider them here. In later chapters they play a prominent role.

Newton's law of motion states that the force $\boldsymbol{F}$ acting on a particle is equal to the rate of change (i.e. the time derivative) of its momentum. For a particle of constant mass, this means that

$$\boldsymbol{F} = mD\boldsymbol{V} = m\boldsymbol{A}. \tag{2.2}$$

Some textbooks still assert, wrongly, that (2.2) *defines* the force. The right hand side of (2.2) depends only on the particle, while the left-hand side may depend on the other particles of the system, and also on the environment. The interpretation of (2.2) as a definition would make it impossible to calculate the motion of a system under given forces, for example.

Gravitational force

If $(\boldsymbol{Z}_i, m_i)$ and $(\boldsymbol{Z}_j, m_j)$ are particles of constant mass m_i and m_j, respectively, the gravitational force exerted by particle i on particle j is

$$\boldsymbol{F}_{ij} = Gm_i m_j(\boldsymbol{Z}_i - \boldsymbol{Z}_j) \mid \boldsymbol{Z}_i - \boldsymbol{Z}_j \mid^{-3}, \tag{2.3}$$

where G, the *Newtonian gravitational constant*, is a universal constant, the same for all pairs of particles. The force $\boldsymbol{F}_{ij}$ is a vector field that depends on the time; to avoid problems with domains of definition, we suppose that $\boldsymbol{Z}_i$ and $\boldsymbol{Z}_j$ are defined on the same time interval. It follows from (2.3) that $\boldsymbol{F}_{ji} = -\boldsymbol{F}_{ij}$. We assume that the paths of the particles do not intersect, so that (2.3) is always well defined.

If more than two particles are present, the gravitational force between each pair of them is given by an equation of the form (2.3). If $\boldsymbol{F}_{ij}$ is the gravitational force exerted by particle i on particle j, for $i, j \in \{1, 2, \ldots, n\}$, $i \neq j$, then the total gravitational force on particle j is $\sum \boldsymbol{F}_{ij}$, where the sum is over all i except $i = j$.

We have defined gravitational force mathematically. In the usual manner, (2.2) sometimes gives the approximate, empirical gravitational force between small material particles. The measured value of the constant is $G = (6.6726 \pm 0.0005) \times 10^{-11}\text{m}^3\text{kg}^{-1}\text{s}^{-2}$[Luther 82].

The gravitational potential

The gravitational force can be written as the gradient of a real function – the Newtonian gravitational potential. Before proving this, we recall some definitions.

Let us suppose that there is a function $\mathcal{U}$ of four variables, such that the force acting at time t on a particle with path $\boldsymbol{Z}$ is

$$\boldsymbol{f}(t) = -\nabla\mathcal{U}(\boldsymbol{Z}(t), t), \tag{2.4}$$

where $\nabla\mathcal{U}$ is the vector field whose r component is $\partial_r\mathcal{U}$, and ∂_r is the partial derivative with respect to the r slot. The minus sign is put into (2.4) so that we can later interpret $\mathcal{U}$ as an energy. A force that satisfies (2.4) for some $\mathcal{U}$ is said to be *lamellar*.

The function $\mathcal{U}$ must be defined and differentiable for all (x_1, x_2, x_3, t) in some domain that contains the path of the particle. The domain must be an open set in $\mathbb{R}^4$, and one may restrict it in some other way, e.g. by requiring it to be star-shaped.[3] One can say that $\nabla\mathcal{U}$ is a vector field on this domain, or that $\nabla\mathcal{U}(\boldsymbol{Z}, .)$

is a vector field that depends on a single variable t (The dot denotes a missing variable t: one defines $\nabla\mathcal{U}(\boldsymbol{Z},\ .)(t) = \nabla\mathcal{U}(\boldsymbol{Z}(t), t)$.)

The gravitational force on a particle of mass m is given by (2.4) with $\mathcal{U} = m\Phi$, where Φ – the (*Newtonian*) *gravitational potential* – is independent of m. We define $-\nabla\Phi$ to be the *gravitational field* (note the sign!). By comparing (2.4) and (2.3), we see that for a system of n particles of mass m_i, $i \in \{1, \ldots, n\}$, the gravitational potential at $(\boldsymbol{x}, t) = (x^1,\ x^2,\ x^3,\ t)$ is

$$\Phi(\boldsymbol{x}, t) = -\sum_i Gm_i \mid \boldsymbol{x} - \boldsymbol{Z}_i(t) \mid^{-1} + \Phi_0(\boldsymbol{x}, t), \tag{2.5}$$

where Φ_0 is the contribution to the gravitational potential due to the environment. When calculating the potential at the position of particle j, one must omit the term with $i = j$ from the sum in (2.5) in order to avoid a divergence. If one wishes to calculate the potential at the position of a body due to the body itself, one must take account of its structure – one cannot regard the body as a particle.

The gravitational forces are unchanged if the gravitational potential Φ is replaced by $\Phi + k$, for any constant k, and we say that the gravitational potential is *arbitrary to the extent of an additive constant.* The significance of this apparently trivial assertion will emerge later.

The gravitational potential is assumed to satisfy Laplace's equation

$$\nabla^2\Phi = 0 \tag{2.6}$$

at all points that do not lie on the path of a particle. This is easy to prove for the first term in (2.5). We assume that it also holds for the Φ_0 term, on the plausible ground that Φ_0 is due to particles in the environment.

One can differentiate (2.5) everywhere, even on the paths of the particles, by using the theory of generalized functions. From Appendix C, eq.(C31), one has $\nabla^2(\mid \boldsymbol{x} \mid^{-1}) = -4\pi\delta^3(\boldsymbol{x})$, where δ^3 is the 3 dimensional Dirac delta function (to check this equation, apply the divergence theorem to $\nabla(\mid \boldsymbol{x} \mid^{-1})$). Og[4]

$$\nabla^2\Phi(\boldsymbol{x}, t) - \nabla^2\Phi_0(\boldsymbol{x}, t) = 4\pi G \sum_i m_i \delta^3(\boldsymbol{x} - \boldsymbol{Z}_i(t)). \tag{2.7}$$

One can interpret the right-hand side (the "source term") as being due to a set of point particles of mass m_i. However, as noted earlier, the association of mathematical point particles with empirical particles can lead to difficulties, and we shall almost always prefer to describe the source of the gravitational field by a mass density – a smooth function ρ of the space and time variables. The integral of ρ over a spatial region at a given instant is the total mass in the region at that instant, and instead of (2.7) one has Poisson's equation:

$$\nabla^2\Phi(\boldsymbol{x},t) = 4\pi G\rho(\boldsymbol{x},t). \tag{2.8}$$

We have assumed in (2.8) that Φ_0 satisfies Laplace's equation: $\nabla^2\Phi_0(\boldsymbol{x},t) = 0$.

Galilean charts

The cartesian chart, whose existence is assumed in the Newtonian theory, is not unique. However, if one chooses a different cartesian chart, one may have to add other terms to the force in (2.2). The additional terms (centrifugal force, Coriolis force, etc.) are proportional to the mass of the particle.

We might hope that a suitable choice of cartesian chart would eliminate from (2.2) all terms proportional to the masses of the particles. Unfortunately, this is not possible: the gravitational forces, which act between every pair of particles, cannot be eliminated. The best that we can do is find a chart in which there are no *non-gravitational* forces that are proportional to the masses. Such a chart is called a *Galilean chart*; a Galilean chart together with a time coordinate is a *Galilean frame*.

A non-gravitational force that is proportional to the mass may be called an *inertial force*. A Galilean frame can therefore be characterized as one in which the inertial forces vanish. Charts in which the inertial forces do not vanish are sometimes said to be *accelerated* because they are accelerated with respect to Galilean frames, and Galilean frames are said to be *unaccelerated*.

To determine experimentally whether a given frame is Galilean is likely to be a difficult task. If we know the distribution of mass in the relevant part of the universe, and accept Newton's law of gravitation, we can calculate the gravitational forces, subtract them from the total forces, and ask whether any of the remainders are proportional to mass. If they are not, then the frame is Galilean.

There are several obvious objections to this program. One must be able to measure all the forces, and must know the distribution of mass, not only in the system but in the rest of the universe. We assume that all such difficulties can be overcome and assert, as a fundamental hypothesis of Newtonian theory, that Galilean frames exist.

One finds from experiment that a cartesian frame in which the distant galaxies are unaccelerated is Galilean. This is rather mysterious. Ernst Mach proposed that the Galilean frames are in fact determined by the matter distribution of the universe (*Mach's hypothesis*), but there is no satisfactory theory that explains how.

There are other problems with the existence of Galilean frames that we shall mention only briefly. We shall see that Newton's theory of gravity is only approximately correct. Any exact theory would almost certainly imply the existence of gravitational waves, and possibly the existence of gravitational fields that cannot be assigned to any material source. In such a theory it would be difficult or impossible to separate gravitational forces from other forces proportional to the mass – and Galilean frames as we have defined them might not exist. However, for most practical purposes they do exist, to a very good approximation, and we shall not worry further about such questions.

Summary

We have now finished our brief review of the Newtonian theory. The only feature of the treatment that the reader may have found unfamiliar is the slightly tedious insistence on the distinction between mathematical and empirical quantities. Physicists commonly avoid making this explicit, and one can argue that they are right – that physics works only through the identification of two quite disparate worlds. We have been forced to emphasize the distinction because, as shown in the next chapter, the conventional, Newtonian identifications of mathematical and empirical quantities are incorrect. We shall have to cut them apart and resew them, like the unfortunate African ladies whose chastity is surgically preserved between pregnancies.

Notes

[1] More precisely, we assume that there is a bijection $\boldsymbol{x}$ (a one-to-one and onto map) from the set $\mathcal{S}$ of spatial points to the set $\mathbb{R}^3$ of ordered triples. We write $\boldsymbol{x} : \mathcal{S} \to \mathbb{R}^3$, and $\boldsymbol{x}(P) = (x^1(P), x^2(P), x^3(P))$. We call the map $\boldsymbol{x}$ a *chart* or *coordinate system*, and $\boldsymbol{x}(P)$ (or the real numbers $x^m(P), m \in \{1, 2, 3\}$) the *coordinates* of the spatial point P in $\boldsymbol{x}$. If talk of bijections and ordered triples is repugnant to the unmathematical reader, he can disregard it. The book contains very little of that kind of language.

[2] In a relativistic theory, one would speak of a *spacetime point* instead of a *point-event*. We use the clumsier designation to emphasize that these points are in 3 dimensional space.

[3] A domain is *star-shaped* if there is a point P in its interior such that every half line with P as endpoint cuts the boundary of the domain in exactly one point.

[4] Og is the acronym for *one gets*. It stands for any of the phrases used by authors when they are tired of writing things out in full: *we have, one shows that, the reader will easily prove that, it is obvious to the meanest intelligence that*, etc.

II Static Gravitational Fields

1. Limitations of the Newtonian theory

We are going to examine the Newtonian theory with a critical eye and show how its defects can be repaired. The approach is minimalist: "If it ain't bust, don't fix it". We shall try to construct a satisfactory theory that is as close as possible to the Newtonian. The modified theory will still have a limited scope. We shall have to determine and, in later chapters, extend its range of validity.

The main limitation of the Newtonian theory is that it applies only to static or quasistatic (i.e. almost static) gravitational fields. From the brief account in Chapter I, it may not be obvious that any such limitation exists; the arguments of the next paragraphs will make it plausible.

The expression for the gravitational force between two particles, eq.I(2.3), depends on the time variable only implicitly, through the positions of the particles. It follows that, if one particle moves, the force on the other changes immediately (at the same instant). We can send signals from one particle to the other by moving the first back and forth a short distance, and measuring the changes in the force on the second. With sufficiently sensitive force detectors, the signals may be sent over great distances, and their speed of propagation can be greater than the speed of light.

In the theory of special relativity, it is shown that signals faster than light give rise to peculiar paradoxes (violations of causality, precognition). However, one cannot assume that gravity is describable within special relativity, and appeals to relativistic principles are consequently dubious. The best that one can argue is that, when the gravitational field is everywhere very weak, the principles of special relativity should be "almost" true – and in particular, the maximum speed of signals should be very close to the speed of light. Obviously such arguments, which are based on some vague idea of physical continuity, cannot be completely convincing. Nevertheless, we provisionally accept them and conclude that the Newtonian theory does not apply to time-varying gravitational fields of the kind that are required for the transmission of signals.

There is a strong analogy between Newtonian gravity and electrostatics that, for historical reasons, is apt to dominate our thinking. The electrostatic field, like the gravitational, is the gradient of a real function (the scalar potential) that satisfies Laplace's equation in regions where there are no sources. We know that electrostatics is a special case of the Maxwell theory of electromagnetism, in which the scalar potential is supplemented by a vector potential, and we might be tempted to generalize Newtonian gravity in a similar manner. In this chapter, we firmly reject such rash assumptions. We consider only static or quasistatic gravitational fields. We show that even here the Newtonian theory is not completely satisfactory, and mend it as invisibly as possible. We make (almost) no hypotheses.[1]

2. Static systems

The mathematical framework on which we build is very similar to that of the Newtonian theory. We assume that there is a cartesian chart $\boldsymbol{x}$ and a time coordinate t. The cartesian chart is not unique: we can shift the spatial origin, or rotate or reflect the spatial axes. In addition, we can perform an affine transformation $t \mapsto t' = at + b$ of the time coordinate, where a and b are constants, $a \neq 0$. We do not however allow transformations $\boldsymbol{x} \mapsto \boldsymbol{x}'$ in which the x'^m are functions of t.

Any cartesian chart $\boldsymbol{x}$ together with a time coordinate t may be called a *Newtonian reference frame* (or *Newtonian frame of reference*), which we write as $(\boldsymbol{x}, t)$. From now on, all reference frames are Newtonian unless stated otherwise. We have at present no way of preferring one Newtonian reference frame over any other, so we take care that our definitions and assumptions are independent of the particular choice.

Newtonian frames play the same role in the modified Newtonian theory (which we call *gravitostatics*) as do Galilean frames in the unmodified theory. We introduce the new name because we shall find that the relationship between empirical and mathematical quantities is different in the two theories.

A spatial point P is said to be *fixed* or *at rest* in the frame $(\boldsymbol{x}, t)$ if its coordinates $\boldsymbol{z} = (z^1, z^2, z^3)$ are independent of t. A spatial point fixed in $(\boldsymbol{x}, t)$ is also fixed in any other Newtonian frame (recall the restrictions on the allowed transformations of $\boldsymbol{x}$ and t); we can therefore say simply that P is a fixed (spatial) point. In later chapters, we shall consider a larger set of *postnewtonian* reference frames, and a point fixed in one frame may not be fixed in another.

A quantity, field, system, or other object is said to be *static* if its description in a reference frame $(\boldsymbol{x}, t)$ is independent of t. For example, the gravitational potential Φ is static if it is a function of the spatial coordinates $\boldsymbol{x}$ but not of t. It is easy to see that the definition does not depend on the choice of reference frame: if something is static in $(\boldsymbol{x}, t)$ then it is static in any other (Newtonian) reference frame $(\boldsymbol{x}', t')$.

In terms of the distinction between mathematical and empirical quantities that was made in Chapter I, our definition of static quantities is mathematical (it refers to a description in a reference frame – which is a mathematical description). If we ask how a static *empirical* quantity is characterized in terms of measurements, we encounter a logical problem. Any measurement requires that something be done – that something change in time. Consequently, in a truly static situation, measurements are impossible. There is no elegant and general way of escaping from this difficulty. In each case, one must try to ensure that the time-dependent effects due to the measurement are small in comparison with the static quantities that are being measured.

In addition to measurements, we have to consider other nonstatic happenings in static systems – we may wish to explore the properties of a static system by sending light signals or small objects from one point to another, for example. Such nonstatic happenings can be described by a set of functions that depend on the time t, and other variables. Because the background system is static, the nonstatic happenings can occur at any time *in the same manner*. More precisely, if one replaces t by $t + K$ in the set of functions that describes them, where K is any constant, then one gets the description of another, exactly similar, physically possible set of happenings. This is sometimes called the "time translation invariance" of physically possible happenings in a static background. We shall use the term *gravitostatics* rather broadly, to mean the theory of the static background field and of possibly nonstatic happenings in it.

General cartesian charts

We shall sometimes consider *scale transformations*, in which one multiplies all three spatial coordinates of a cartesian chart $\boldsymbol{x}$ by a strictly positive constant, which we write as σ^{-1}. The resulting chart $\boldsymbol{x}' = \sigma^{-1}\boldsymbol{x}$ is a general cartesian chart, as defined in Chapter I, eq.I(2.1). The Euclidean distance between the spatial points P and Q is

$$\overline{PQ} = |\boldsymbol{x}(P) - \boldsymbol{x}(Q)| = \sigma|\boldsymbol{x}'(P) - \boldsymbol{x}'(Q)|. \tag{2.1}$$

Although we have spoken of *the* Euclidean distance, it is not mathematically unique. One can equally well define another Euclidean distance by

$$(\overline{PQ})' = |\boldsymbol{x}'(P) - \boldsymbol{x}'(Q)| = \sigma^{-1}\overline{PQ}, \tag{2.2}$$

and one easily checks that the axioms of Euclidean geometry are satisfied with the second definition if they are with the first. There is not much point to this in Newtonian theory, where Euclidean distance is identified with empirical distance, and is uniquely determined once the empirical unit of length is chosen. However, we shall find it useful later, when considering spaces that are non-Euclidean. By performing an appropriate scale transformation, one can ensure that the Euclidean distance (2.2) is approximately equal to the empirical distance in a restricted spatial region where the potential is almost constant.

3. Gravitational red shift

We must now consider more carefully how the time variable of a Newtonian frame is related to time measured by clocks. This will lead us to ask how the readings of clocks at different points are related, and how signals travel from one point to another in a static gravitational field.

We take the spatial point P to be at rest with constant spatial coordinates $\boldsymbol{z} = (z^1, z^2, z^3)$ in the Newtonian frame $(\boldsymbol{x}, t)$. As in the Newtonian theory, we assume that P can sometimes be associated with a clock. The clock must be a small material object, so that it can be associated more or less unambiguously with P, and we say that the clock is *fixed at* P, or is *at rest in* $(\boldsymbol{x}, t)$ with constant spatial coordinates $\boldsymbol{z}$.

Real clocks are of many kinds and have numerous defects, which we shall not discuss. Our clocks are all of the same kind, and are ideal: they are unaffected by temperature, voltage surges, and vagrant cockroaches, and keep good time when compared with the standard, atomic clocks of national laboratories.

We assume, as in the Newtonian theory, that the time T measured by a clock at P can be associated with the time coordinate t, but we do not now assert that their values must agree. Let us suppose that the clock C, which is fixed at P, reads T when the time coordinate is t, and T' when the time coordinate is t'. (More ponderously, we may say that the clock reads T when it is associated with the point-event (P, t); more simply, that it reads T at t, etc.). In the Newtonian theory, one *assumes* that $T' - T = t' - t$, or that $T = t$ with a suitable choice of origin. This equation looks peculiar: the left-hand side is a measurable, empirical quantity, but the right-hand side is mathematical. It asserts that the values of the mathematical quantity are the same as the measured values of the empirical quantity. (It might be better to write $T \cong t$ to indicate that T and t are quantities of different kinds.)

In such circumstances, physicists like to "identify" the mathematical and empirical quantities, and speak of a single *physical* quantity with mathematical and empirical aspects. They make sibylline definitions: *a variable is physical if it is mathematically well defined and directly measurable*. Imprecise and no doubt deplorable, but that is the way physics is done.

We shall exercise caution. We are not going to assert dogmatically that T and t can be identified in the Newtonian manner. To understand their relationship, we must consider a more complex situation. We suppose that two clocks, C_P and C_Q, are fixed at the points P and Q, respectively. Signals are emitted from C_P at times T_P and T'_P, and arrive at C_Q at times T_Q and T'_Q, respectively, where T_P and T'_P are measured by C_P, and T_Q and T'_Q by C_Q. The times T_P, T'_P, T_Q, T'_Q, correspond to the values of the time coordinate t_P, t'_P, t_Q, t'_Q, respectively. It does not matter what signals are sent, so long as they are of exactly the same kind. For definiteness, one might think of them as light signals, both sent through vacuum along the same spatial path. The signals are, of course, not static, but they move in a static background, and their descriptions in the chart $\boldsymbol{x}$ are therefore exactly similar (cf. the remarks at the end of Section 2). In particular, the coordinate times between emission and arrival are the same: $t_Q - t_P = t'_Q - t'_P$, or equivalently

$$t'_Q - t_Q = t'_P - t_P. \tag{3.1}$$

The times measured by the clocks satisfy an equation slightly different from (3.1). If Φ_P and Φ_Q are the values of the gravitational potential at the positions of C_P and C_Q, respectively, and c is the speed of light, we find from experiment that

$$T'_Q - T_Q = (T'_P - T_P)\left[1 + (\Phi_Q - \Phi_P)c^{-2}\right]. \tag{3.2}$$

If $\Phi_Q > \Phi_P$, this implies that the measured time interval between the arrival of the two signals at C_Q is greater than the time interval between their emission from C_P.

Instead of sending two discrete signals, we may think of transmitting a periodic wave from one clock to the other. We can then regard $T_Q = T'_Q - T_Q$ as the measured period of the wave at C_Q and $T_P = T'_P - T_P$ as that at C_P, and (3.2) becomes

$$T_Q = T_P\left[1 + (\Phi_Q - \Phi_P)c^{-2}\right]. \tag{3.3}$$

We have implicitly assumed in the derivation of (3.2) and (3.3) that $T'_P > T_P$ and $T'_Q > T_Q$. This is possible only if C_P and C_Q agree in their conventions about past and future (C_P must not "run backwards" with respect to C_Q).

In the case of a light wave, one has $T_Q > T_P$ when $\Phi_Q > \Phi_P$, and the received light is "redder" than that transmitted. Because the early measurements were on light from the Sun or stars, which appears anomalously red to astronomers on Earth, the effect is usually called the *gravitational red shift*. The first laboratory experiments used Mössbauer radiation – which consists of γ rays of very well defined frequency [Pound 65]. More precise measurements have used rockets [Vessot 80].

Exercise 3.1 A periodic wave is transmitted from P to Q, as described above. Show that, if the source at P goes through m periods between the coordinate times t_1 and t_2, then the light arriving at the receiver at Q must go through the same number of periods between t_1 and t_2. [If it did not, the phase change along the path of the light (the number of wave peaks between the source and receiver) would not be the same at t_1 and t_2.] ■ (The symbol ■ denotes the end of an exercise or proof.)

The gravitational red shift experiments show that we cannot in general identify the time coordinate t with time as measured by stationary, ideal clocks. It is true that the term $(\Phi_Q - \Phi_P)c^{-2}$ is usually very small – between the surface of the Sun and the Earth it is of order 10^{-6}, for example – but it is certainly not negligible.

Eq.(3.3) is empirical. We must now ask whether we can derive it from general, physically plausible hypotheses. The rather astonishing answer is that we can derive it without any hypotheses at all – or rather, without any *new* hypotheses. We shall show that it follows from the standard assumption of the Newtonian theory that the gravitational potential is arbitrary to the extent of an additive constant, together with some other well established laws. All that we have to admit is the possibility that physical quantities – in this case the period of a wave – may depend on the gravitational potential. We shall prove that the periods T_P and T_Q are related by the equation

$$T_Q = T_P \exp \tau \Big[(\Phi_Q - \Phi_P)c^{-2}\Big], \tag{3.4}$$

where τ is a universal constant. Eq.(3.4) is equivalent to (3.3) for small differences of potential, provided that τ is close to 1. We later show that in fact $\tau = 1$.

4. Arbitrariness of Φ

In the Newtonian theory, it is assumed – or is trivially evident – that the gravitational potential Φ is arbitrary to the extent of an additive constant. We reassert this as a formal hypothesis of the theory. More precisely, it states that the value of any measurable quantity is unchanged when the function Φ is replaced by $\Phi + K$, for any constant K. It follows that no measurement can determine the value of Φ at a point (or, more precisely, in a small neighbourhood). One can measure only differences between the values of Φ at different points.

An immediate consequence of the hypothesis is that the measured values of universal constants are independent of Φ. (By *universal constants*, we mean such things as the speed of light in vacuum c, the electronic charge e, and Planck's constant h.) For if some universal constant ξ were a non-constant function of Φ, one could determine the value of Φ in a neighbourhood by simply measuring ξ there.

The preceding argument applies to any well defined physical quantity, and not just to universal constants. For example, the frequency of a given spectral line, emitted by an atom and measured by a clock that are both at rest near the spatial point P, is the same as the frequency of the same line, emitted by an exactly similar atom and measured by a clock that are both at rest near the spatial point Q. Similarly, the ratio of the measured lengths of two small objects near P is the same as their ratio near Q – provided that the objects can be transported "without change" from P to Q. More generally, the measured value of any well defined, dimensionless quantity is the same at P and at Q.

Eq.(3.4), which shows the dependence of the period of a wave on the gravitational potential Φ, is a special case of a more general equation, which we now derive. We define a *local* quantity to be one that refers to a spatial region where Φ does not change appreciably from point to point. We consider an empirical, local quantity $\mathcal{Q}$ that is associated with a mathematical quantity q, in the manner that we have previously discussed. (To make the argument less abstract, one can, for example, think of $\mathcal{Q}$ as being the measured period of a wave, and q as being the difference between the time coordinate t at successive peaks, as in the last section.) We are going to consider local quantities in different spatial regions; we therefore write $\mathcal{Q}_P$ and q_P, instead of $\mathcal{Q}$ and q, where P is any point in the spatial region to which the quantities refer. The gravitational potential in the region is Φ_P (by assumption, it does not vary appreciably in the region and is independent of the particular choice of P), and we define $R_P = \mathcal{Q}_P/q_P$.

The Newtonian assumption is that empirical and mathematical quantities can be identified, so that $R_P = 1$. We have seen that this is inconsistent with experiment in at least one case, and we wish to consider the more general possibility that R_P depends on the potential. To be more precise, let $\mathcal{Q}_Q$ be a local quantity exactly similar to $\mathcal{Q}_P$ but in a region labelled by a point Q. (We shall be logically loose and say that $\mathcal{Q}_P$ and $\mathcal{Q}_Q$ are the *same* local quantity in the regions labelled by P and Q.) We define $R_Q = \mathcal{Q}_Q/q_Q$, where q_Q is the associated mathematical quantity, and assume that for some function f

$$R_Q/R_P = f(\Phi_Q, \Phi_P). \tag{4.1}$$

Writing (4.1) for regions labelled by points S and Q, and S and P, og $R_S/R_Q = f(\Phi_S, \Phi_Q), R_S/R_P = f(\Phi_S, \Phi_P)$, and hence

$$f(\Phi_S, \Phi_P) = f(\Phi_S, \Phi_Q) f(\Phi_Q, \Phi_P). \tag{4.2}$$

We now use the assumption that the value of any measurable quantity is unchanged when Φ is replaced by $\Phi + K$, for any constant K. We note that the ratio $R_Q/R_P = (\mathcal{Q}_Q/q_Q)/(\mathcal{Q}_P/q_P) = (\mathcal{Q}_Q/\mathcal{Q}_P)(q_P/q_Q)$ is measurable if the ratio q_P/q_Q of the two mathematical *quantities* can be determined. (The determination of q_P/q_Q may be by a theoretical argument, or by an arbitrary choice – *provided that the choice is not frame dependent.*) It follows that $f(\Phi_Q, \Phi_P)$ is a measurable quantity whose value is unchanged when Φ is replaced by $\Phi + K$:

$$f(\Phi_Q + K, \Phi_P + K) = f(\Phi_Q, \Phi_P), \tag{4.3}$$

for all K.

We can solve (4.2) and (4.3) to find the function f. Choosing $K = -\Phi_P$ in (4.3), og $f(\Phi_Q, \Phi_P) = f(\Phi_Q - \Phi_P, 0)$. We define a function u by $u(z) = f(z, 0)$, and assume that u is defined and differentiable for all real z in some open interval that includes $z = 0$. Eq.(4.2) then becomes $u(z+w) = u(z)u(w)$. The solution of this equation that satisfies $u(0) = 1$ (corresponding to $f(\Phi_P, \Phi_P) = R_P/R_P = 1$) is $u(z) = \exp(\alpha c^{-2} z)$, where α is a constant and c is the speed of light – the factor c^{-2} is inserted to make α dimensionless. We have therefore proved that

$$R_Q/R_P = f(\Phi_Q, \Phi_P) = f(\Phi_Q - \Phi_P, 0) = \exp\left[\alpha(\Phi_Q - \Phi_P)c^{-2}\right]. \tag{4.4}$$

Exercise 4.1 Solve the equation $u(z+w) = u(z)u(w)$ under the assumptions given above ($u(0) = 1$ and differentiability). ■

We assumed in the derivation of (4.4) that the ratio q_P/q_Q can be determined – by a theoretical argument or an arbitrary choice that is not frame dependent. The simpler assumption that q_P can be determined is not possible because q_P usually depends on the choice of frame, and the theory does not include a prescription for a unique Newtonian frame.

To show that (4.4) reduces to (3.4) as a special case, we write $\alpha = \tau, R_P = T_P/(t'_P - t_P)$, and $R_Q = T_Q/(t'_Q - t_Q)$, and note that $(t'_P - t_P)/(t'_Q - t_Q) = 1$, from (3.1). In this case we have determined $q_P/q_Q = (t'_P - t_P)/(t'_Q - t_Q)$ from a theoretical argument – from the assumption that the wave moves in a static background (which implies that its period is the same at P and Q, when measured in terms of the coordinate t).

We can interpret (4.4) in a different way. We again take the special case $R_P = T_P/(t'_P - t_P)$ and $R_Q = T_Q/(t'_Q - t_Q)$, but now interpret T_P and T_Q as the periods of two exactly similar systems at P and Q, respectively (e.g. the periods of the same spectral line of hydrogen atoms at rest near P and Q). We then have $T_P = T_Q$ (cf. the discussion near the beginning of this section), and (4.4) (with $\alpha = \tau$) gives

$$t'_P - t_P = (t'_Q - t_Q)\exp\left[\tau(\Phi_Q - \Phi_P)c^{-2}\right]. \tag{4.5}$$

This shows how the periods of exactly similar systems differ from place to place when expressed in terms of the time coordinate of a Newtonian reference frame.

For brevity, we say that a time interval expressed in terms of the time coordinate of a Newtonian reference frame is a *Newtonian time interval*, or a *time interval in Newtonian units*. Similarly, we may refer to the distance $|\boldsymbol{z}(P) - \boldsymbol{z}(Q)|$ as the *Newtonian distance* between the points P and Q, or the *distance in Newtonian units*, where $\boldsymbol{z}(P)$ and $\boldsymbol{z}(Q)$ are the coordinates of P and Q in a Newtonian chart. We define Newtonian units of velocity, acceleration, etc., in the obvious way. One must remember when using Newtonian units that they depend on the choice of Newtonian frame. Scale transformations of the space or time coordinates give different Newtonian units (cf. the discussion at the end of Section 2).

Sometimes we may speak of quantities *measured* in Newtonian units. This is probably harmless, but it is another example of how mathematical and empirical terminology contaminate each other. We have to keep in mind that real measurements, with apparatus, are in terms of *empirical* units. If we wish to emphasize that a quantity is measured in empirical units, we may put a suffix E on it; similarly, we may denote quantities in Newtonian units with a suffix N. In (4.5), for example, we might write $t'_P - t_P = T_{NP}$ as the period of the system at P in Newtonian units, and T_{EP} as the experimentally measured period.

5. Φ dependence of length

Eq.(4.4) can be used to find the relationship between lengths in different regions, although the restriction to local quantities means that we must consider regions where Φ does not change appreciably. We may take L_P, for example, to be the length of a small measuring rod in a region where $\Phi \approx \Phi_P$. We take l_P to be the Newtonian (or Euclidean) distance between the points P' and P'' that are associated with the ends of the rod. (For brevity, we say that l_P is the Euclidean distance that corresponds to L_P.) If the coordinates of the points P' and P'' in the cartesian chart $\boldsymbol{x}$ are $\boldsymbol{z}'$ and $\boldsymbol{z}''$, we have $l_P = |\boldsymbol{z}' - \boldsymbol{z}''|$. Similarly, in a region where $\Phi \approx \Phi_Q$, we take L_Q to be the measured length of a rod, and $l_Q = |\boldsymbol{w}' - \boldsymbol{w}''|$ to be the Euclidean distance between the points Q' and Q'' associated with its ends, whose coordinates in $\boldsymbol{x}$ are $\boldsymbol{w}'$ and $\boldsymbol{w}''$.

In (4.4) we now substitute $R_P = L_P/l_P$, and $R_Q = L_Q/l_Q$. We make the restriction that

$$l_P = l_Q, \tag{5.1}$$

which is frame independent, and therefore permissible. Replacing the constant α by λ (for λength), og

$$L_Q = L_P \exp\left[\lambda(\Phi_Q - \Phi_P)c^{-2}\right]. \tag{5.2}$$

The equation $l_P = l_Q$ means that the empirical lengths L_P and L_Q correspond to the *same* Euclidean distance. We recall that we made a correspondence between time intervals measured in two different regions in a very similar manner: instead of (5.1), we used the equality of the coordinate time differences, eq.(3.1).

If L_P and L_Q are the measured lengths of two exactly similar measuring rods in the regions labelled by P and Q (or of the same rod, transported without distortion from one region to the other), then instead of (5.1) we have $L_P = L_Q$, and (4.4) becomes

$$l_P = l_Q \exp\left[\lambda(\Phi_Q - \Phi_P)c^{-2}\right]. \tag{5.3}$$

This shows how the Newtonian length corresponding to a given empirical length differs from point to point.

The (mathematical) statement that space is Euclidean is equivalent to saying that there is a chart $\boldsymbol{x}$ in which the distance between any pair of points P and Q is $\overline{PQ} = |\boldsymbol{x}(P) - \boldsymbol{x}(Q)|$ as in (2.1). It follows from (5.1) and (5.2) that if $\lambda = 0$, then the measured distance between two particles associated with P and Q is equal to $\overline{PQ}$, and the geometry of physical space is Euclidean, just as in Newtonian theory. This means, for example, that the measured lengths of the sides of a triangle satisfy the usual Euclidean relationships. If however $\lambda \neq 0$ and Φ is not constant, then the geometry of physical space is not Euclidean. In the terminology of the last section, space is Euclidean when lengths are measured in terms of Newtonian units, but not when measured in terms of empirical units.[2] The value of λ might be found by careful measurements of the sides of a triangle, for example – although this is very hard to do in practice, and we shall determine λ in another way.

6. Φ dependence of mass

Eq.(4.4) can also be used to find the relationship between masses in different regions. The mass of a small material object can be considered to be a local quantity provided that the mass is not so large or concentrated that Φ changes appreciably over the volume of the object. In a region where $\Phi \approx \Phi_P$, we write the empirical mass of the object as M_P and the associated mathematical mass as m_P, and we take $R_P = M_P/m_P$, $R_Q = M_Q/m_Q$ in (4.4). (Rather than *mathematical* mass, we shall usually say *Newtonian* mass, in the same way that we spoke of Newtonian time interval and distance.)

If M_P and M_Q are the empirical masses of two exactly similar objects (two protons, or two standard kilogram masses, for example), we have $M_P = M_Q$, and (4.4) gives

$$m_P = m_Q \exp\left[\mu(\Phi_Q - \Phi_P)c^{-2}\right], \tag{6.1}$$

for some constant μ.

Eq.(6.1) shows how the Newtonian mass of an object of constant empirical mass depends on the gravitational potential. The usual assumption in Newtonian mechanics is that the mass m of a particle is constant along its path – which would imply that $\mu = 0$. However, we recall that in special relativity the mass depends on the speed of the particle, and there seems to be no good reason why it should not also depend on the potential. Instead of speculating, we shall defer discussion of the value of μ until we have considered particle mechanics in the next chapter.

7. Φ dependence of local quantities

The arguments of Section 4 apply to any local quantity Q, and in particular to any Q that is dimensionless. As at the beginning of Section 4, we assume that any well-defined, dimensionless, empirical local quantity is independent of Φ (if it were not, one could determine Φ by measuring the quantity – which would contradict the assumption that Φ is arbitrary to the extent of an additive constant). We then define the associated Newtonian quantity q by $q_P = \mathcal{Q}_P$ for all P, which implies that $\alpha = 0$ in (4.4).

We are going to consider only quantities that have well defined physical dimensions. We take the basic dimensions to be length, mass, and time, and denote the dimensions of the quantity $\mathcal{Q}$ by $[\mathcal{Q}] = [\mathcal{L}^\xi \mathcal{M}^\eta \mathcal{T}^\zeta]$, where ξ, η, and ζ are real. For a speed v, for example, one has $\xi = 1$, $\eta = 0$, and $\zeta = -1$, and $[v] = [\mathcal{L}\mathcal{T}^{-1}]$.

It is easy to show that local quantities with the same dimensions have the same Φ dependence (the same value of α in (4.4)). For suppose that $\mathcal{Q}_1$ and $\mathcal{Q}_2$ are local quantities with the same dimensions for which the values of α are α_1 and α_2, respectively. Then $\mathcal{Q} = \mathcal{Q}_1/\mathcal{Q}_2$ is a local quantity with zero dimensions for which α has the value $\alpha_1 - \alpha_2$. But from our previous argument, $\alpha = 0$ for a quantity with zero dimensions, and hence $\alpha_1 = \alpha_2$.

Summary of results

For convenience of reference, we collect together some of the preceding results. To save much writing of c^{-2}, we define the *dimensionless potential* $\phi = \Phi c^{-2}$, where c is the measured value of the speed of light in vacuum (a universal constant, and independent of Φ). We may speak of either Φ or ϕ as the potential, if no ambiguity arises.

Let us suppose that a local quantity has the empirical value $\mathcal{Q}_P$ in a region where the potential is ϕ_P, and an exactly similar local quantity has the same value $\mathcal{Q}_Q = \mathcal{Q}_P$ in a region where the potential is ϕ_Q. The associated Newtonian (or mathematical) quantities have the values q_P and q_Q, where

$$q_P = q_Q \exp[\alpha(\phi_Q - \phi_P)]. \tag{7.1}$$

One can streamline the notation in (7.1) by dropping the suffixes P, and replacing Q by 0:

$$q = q_0 \exp\left[\alpha(\phi_0 - \phi)\right], \tag{7.2}$$

where q is the quantity in a region where the potential is ϕ, and q_0 the quantity in a region where the potential is ϕ_0. One may regard q and ϕ as functions whose arguments are the spatial points or the coordinates of a Newtonian frame, as convenient.

For the special cases when $\mathcal{Q}$ has the dimensions of time, length, or mass, eq.(7.2) becomes (cf. (4.5), (5.3), (6.1))

$$\mathrm{t} = \mathrm{t}_0 \exp\left[\tau(\phi_0 - \phi)\right], \tag{7.3}$$

$$l = l_0 \exp\left[\lambda(\phi_0 - \phi)\right], \tag{7.4}$$

$$m = m_0 \exp\left[\mu(\phi_0 - \phi)\right]. \tag{7.5}$$

In (7.3), we have written t instead of t to avoid confusion with the time coordinate. In the general case, when $[\mathcal{Q}] = [\mathcal{L}^{\xi}\mathcal{M}^{\eta}\mathcal{T}^{\zeta}]$, the value of α in (7.1) and (7.2) is given by

$$\alpha = \xi\lambda + \eta\mu + \zeta\tau. \tag{7.6}$$

Exercise 7.1 Use (7.3) to derive (3.4). [Note that the Newtonian period of a wave is constant from point to point, and remember that the Newtonian time intervals in (7.3) correspond to *equal* empirical time intervals at different points. Is it true that a local Newtonian quantity is proportional to the associated empirical quantity? That is, if t corresponds to T, does βt correspond to βT for every (small) constant β]? ■

Instead of denoting an empirical quantity by an upper-case letter and the associated Newtonian quantity by the corresponding lower-case letter, we shall normally use subscripts E and N, as at the end of Section 4. For example, we may write $\mathcal{Q} = q_E$ and $q = q_N$, or $\mathcal{Q} = \mathcal{Q}_E$ and $q = \mathcal{Q}_N$. To prevent a surfeit of subscripts, we make explicit a convention that has already been used implicitly: universal constants written without subscripts E or N are to be interpreted as empirical quantities unless stated otherwise (c is to be interpreted as c_E, h as h_E, etc.). Other quantities written without subscripts E or N are to be interpreted as Newtonian quantities unless stated otherwise (the velocity $\boldsymbol{v}$ of a particle is to be interpreted as $\boldsymbol{v}_N$, etc.). For these conventions to be workable, we must not use E or N to denote a spatial point!

Eqs.(7.2)–(7.5) hold for any Newtonian frame. One can always choose a particular frame (and a particular definition of the Newtonian masses) so that the value of q when the potential is ϕ_0 is the same as the empirical value: $q_o = q_E$. We shall call such a frame a ϕ_0 *frame*. In a ϕ_0 frame, we may write (7.5), for example, as

$$m = m_N = m_E \exp\left[\mu(\phi_0 - \phi)\right]. \tag{7.7}$$

For the speed of light, we have $\xi = 1$, $\eta = 0$, $\zeta = -1$ in (7.6). In (7.2) we write $q = q_N = c_N$, $c_0 = c_E = c$ (using the new conventions), and get

$$c_N = c \exp\Big[(\lambda - \tau)(\phi_0 - \phi)\Big]. \tag{7.8}$$

8. The metric

In Chapter I we made the distinction between empirical and mathematical quantities – the empirical time T measured by a clock, and the mathematical time coordinate t, etc. The motive was to prepare ourselves for the rejection of the traditional identifications of these quantities. However, we emphasized that one cannot separate them cleanly and completely, and even suggested that physics is made possible only by confusing the two.

We have discovered new relationships between empirical and mathematical (or Newtonian) quantities that replace the unsatisfactory, traditional identifications. If $T_N = \Delta t$ is the difference between two values of the time coordinate, and T_E is the associated time interval measured by an ideal clock, all in a region Ω where the gravitational potential is almost constant, then instead of the traditional relationship $T_E = T_N$, one has $T_E = T_N e^{\tau\psi}$ from (7.3), where $\psi = \phi - \phi_0$, and the Newtonian frame and ϕ_0 are chosen so that $T_0 = T_E$ (we write $e^{\tau\psi}$ instead of $\exp \tau\psi$ for compactness, and ψ is to be evaluated at some point of Ω). Similarly, if $l_N = |\Delta \boldsymbol{x}|$ is the Newtonian distance between neigbouring points whose coordinates in a Newtonian frame are $\boldsymbol{x}$ and $\boldsymbol{x} + \Delta\boldsymbol{x}$, and l_E is the associated empirical distance, one has $l_E = l_N e^{\lambda\psi}$ from (7.4).

It may seem that, although we are wiser and tireder after all our exploring, the only real difference is that we now identify T_E with $T_N e^{\tau\psi}$ rather than T_N, and similarly for other quantities. This is not far from the truth, so long as we restrict ourselves to a ϕ_0 frame in which Newtonian quantities are the same as empirical ones in regions where $\phi \approx \phi_0$. The distinctions between mathematical and empirical quantities will come to the fore again later, when we are forced to consider other charts. In anticipation, we are now going to discuss the representation of empirical lengths and times in terms of an arbitrary chart. *If you wish, you can postpone reading the rest of this section until it is needed in Chapter V.*

We first rewrite our previous expression for l_E. To avoid square roots, we consider l_E^2 instead of l_E. We choose a ϕ_0 frame with $l_0 = l_E$ and $l^2 = (\Delta x^1)^2 +$

$(\Delta x^2)^2 + (\Delta x^3)^2$, and (7.4) becomes

$$l_E^2 = e^{2\lambda\psi}\Big[(\Delta x^1)^2 + (\Delta x^2)^2 + (\Delta x^3)^2\Big] = \sum_{m,n} g_{mn}\Delta x^m \Delta x^n, \tag{8.1}$$

where

$$g_{mn} = \delta_{mn} e^{2\lambda\psi}, \tag{8.2}$$

and we make the convention that lower-case Latin indices have the range $\{1,\ 2,\ 3\}$. The symbol δ_{mn} is the *Kronecker delta*: $\delta_{mn} = 0$ when $m \neq n$ and $\delta_{mn} = 1$ when $m = n$. The g_{mn} are functions of the space and time coordinates which, like ψ, may be evaluated at any point in the neighbourhood.

If we introduce new spatial coordinates $\boldsymbol{x}'$ that are smooth functions of the coordinates $\boldsymbol{x}$ (but independent of t) then, neglecting terms of second degree, og $\Delta x^m = \sum_n (\partial x^m/\partial x'^n)\Delta x'^n$, and (8.1) becomes

$$l_E^2 = \sum_{m,n} g'_{mn}\Delta x'^m \Delta x'^n, \tag{8.3}$$

with

$$g'_{mn}(\boldsymbol{x}', t) = \sum_{r,s} (\partial x^r/\partial x'^m)(\partial x^s/\partial x'^n) g_{rs}(\boldsymbol{x}, t). \tag{8.4}$$

The left-hand side of (8.3) is the square of the empirical length, but the coordinates $\boldsymbol{x}'$ have, in general, no simple physical significance. Because (8.1) is of the same form as (8.3), we say that it is *covariant* (i.e. unchanged in form) *with respect to transformations of the spatial coordinates*, and that l_E^2 is *invariant* with respect to such transformations. The functions g'_{mn} may however be very different from the g_{mn}.

We say that the g_{mn} are the *components of the spatial metric in the frame* $\boldsymbol{x}$ (and similarly the g'_{mn} are the components in the frame $\boldsymbol{x}'$). This is a very simple and traditional way of introducing the metric. More elegant treatments can be found in the geometry books.

In an exactly similar manner, we can introduce a new time coordinate t' that is a smooth function of the coordinate t (but independent of the spatial coordinates $\boldsymbol{x}$). Neglecting terms of second degree, og $T_N = \Delta t = (dt/dt')\Delta t'$, and (7.3) gives

$$T_E = T_N e^{\tau\psi} = (dt/dt')e^{\tau\psi}\Delta t'. \tag{8.5}$$

More generally, we can introduce new space and time coordinates that are functions of the old space *and* time coordinates. Instead of t and t', we use coordinates $x^0 = ct$ and $x'^0 = ct'$, so that space and time coordinates have the same dimension $[\mathcal{L}]$. We write $x = (x^0, \boldsymbol{x}) = (x^0,\ x^1,\ x^2,\ x^3)$, and make the convention that lower-case Greek indices have the range $\{0,\ 1,\ 2,\ 3\}$ and lower-case Latin the range $\{1,\ 2,\ 3\}$. We assume that the x'^{μ} are smooth functions of x.

Eqs.(8.3) and (8.5) (with dt/dt' replaced by $\partial x^0/\partial x'^0$) do not hold for the more general coordinate transformations. However, we can write the quantity $l_E^2 - c^2T_E^2$ in the form

$$l_E^2 - c^2T_E^2 = \sum_{\mu,\nu} g_{\mu\nu}\Delta x^{\mu}\Delta x^{\nu}, \tag{8.6}$$

where g_{mn} are given by (8.2), $g_{00} = -e^{2\tau\psi}$, and $g_{m0} = g_{0m} = 0$. More succinctly, the $g_{\mu\nu}$ are defined by (8.2) and

$$g_{\mu 0} = g_{0\mu} = -\delta_{\mu 0}e^{2\tau\psi}. \tag{8.7}$$

An argument exactly like that which led to (8.3) then shows that

$$l_E^2 - c^2T_E^2 = \sum_{\mu,\nu} g'_{\mu\nu}\Delta x'^{\mu}\Delta x'^{\nu}, \tag{8.8}$$

where

$$g'_{\mu\nu}(\boldsymbol{x}', t) = \sum_{\rho,\sigma}(\partial x^{\rho}/\partial x'^{\mu})(\partial x^{\sigma}/\partial x'^{\nu})g_{\rho\sigma}(\boldsymbol{x}, t). \tag{8.9}$$

We say that the $g_{\mu\nu}$ are the *components of the spacetime metric in the chart* x (and similarly the $g'_{\mu\nu}$ are the components in x'). Since we are now regarding x and x' as spacetime charts, we no longer have to talk about frames (recall that a Newtonian frame was a cartesian chart together with a time coordinate). Because eq.(8.6) has the same form as (8.8), we say that it is *covariant with respect to transformations of the spacetime coordinates*. The quantity $l_E^2 - c^2T_E^2$ is *invariant* with respect to such transformations. Less precisely, we say that a quantity is *physical* if it is mathematically invariant (under an appropriate group of transformations) and directly measurable. The invariant $\sum_{\mu,\nu} g_{\mu\nu}\Delta x^{\mu}\Delta x^{\nu}$ is a physical quantity; the coordinate difference Δx^{ν} is not.

You may have already met (8.6) in special relativity, where the $g_{\mu\nu}$ have the special values $g_{\mu\nu} = \eta_{\mu\nu}$, with $\eta_{mn} = \delta_{mn}, \eta_{\mu 0} = -\delta_{\mu 0}$, and the Δx^{μ} are arbitrary (i.e. one is not restricted to small values of $|\Delta x^{\mu}|$). In this case, we replace the Δx^{μ} by $x^{\mu} - y^{\mu}$, where x and y are the coordinates of arbitrary spacetime points, and define the invariant function I (the *interval function*) by

$$I(x, y) = \sum_{\mu,\nu} \eta_{\mu\nu}(x^{\mu} - y^{\mu})(x^{\nu} - y^{\nu}). \tag{8.10}$$

The interval function is invariant under a coordinate transformation iff the transformation is a Poincaré transformation. A proof is given in Appendix B.

Notes

[1] *Hypothesis non fingo* (*I make no hypothesis*) said Newton, and meant that he was not assuming anything about the physical mechanism underlying his laws. The scope of the word has changed, and we now call the laws themselves *hypotheses*. A law is a hypothesis that we think is well established.

[2] Poincaré pointed out ([Poincaré 52], Chap. V) that it makes no sense to assert without qualification that the geometry of physical space is, or is not, Euclidean. We are free, within wide limits, to change either the geometry or the rules for its interpretation (the empirical definition of *length*); only the combination of the two is physically significant.

III Particle Mechanics

1. Particle mechanics in special relativity

In this chapter, we are concerned with the motion of particles (small material bodies) in a static, or quasistatic, gravitational field. We do not want to restrict ourselves to particles whose speeds are small compared with the speed of light; we therefore construct the theory by generalizing special relativistic dynamics.

The special relativity that we need is summarized in the next few paragraphs. No previous acquaintance with the subject is required, but if the reader feels uneasy about accepting unproved equations, he[1] should read Appendix A (*Kindergarten relativity*).

In special relativity, there are preferred, spacetime coordinate systems called *inertial charts*, with three cartesian spatial coordinates and one time coordinate. They resemble the Galilean frames of Newtonian theory, and the Newtonian frames of gravitostatics that were defined in Chapter II. Inertial charts are however related by Lorentz or Poincaré transformations – which are usually not the same as the transformations that relate Galilean frames.

We are not at present concerned with time-dependent transformations between charts – the Newtonian frames allow only shifts of origin, rotation or reflection of spatial axes, and possibly scale changes. Consequently, we find that we can establish a correspondence between suitable Newtonian frames and inertial charts, such that one reduces to the other in the limit of vanishing gravitational field.

The path of a particle in special relativity can be described, just as in Newtonian theory, by a map $\boldsymbol{Z} : \mathbb{I} \to \mathbb{R}^3$, where $\mathbb{I}$ is an open interval in $\mathbb{R}$, and $\boldsymbol{Z}(t) = (Z^1(t), Z^2(t), Z^3(t))$ are the spatial coordinates of the particle at time t. The particle is associated with a constant m, which is called its *proper mass*. The *3 velocity* and *3 acceleration* of the particle are $\boldsymbol{V} = D\boldsymbol{Z}$ and $\boldsymbol{A} = D\boldsymbol{V}$, respectively: they are the velocity and acceleration of Newtonian theory under different names. The *(mechanical) momentum* is $m\gamma\boldsymbol{V}$, where $\gamma = (1 - |\boldsymbol{V}|^2 c^{-2})^{-1/2}$, and the *energy* is $E = m\gamma c^2$. For a free particle (i.e. one on which no force acts), the momentum and energy are independent of t, and we say they are constants of the motion.

The motion of a particle, in Newtonian or relativistic mechanics, can be described by a *Lagrangian*. This is a function L of the position coordinates $\boldsymbol{Z} = (Z^1, Z^2, Z^3)$, the velocity components $\boldsymbol{V} = (V^1, V^2, V^3)$, and the time coordinate t. The partial derivative of L with respect to Z^i gives the i component of the force $\boldsymbol{F}$ on the particle, and the partial derivative with respect to V^i gives the i component of the *canonical momentum* $\boldsymbol{P}$, for $i \in \{1, 2, , 3\}$:

$$F^i(t) = \partial_i L(\boldsymbol{Z}(t), \boldsymbol{V}(t), t), \qquad P^i(t) = \partial_{i+3} L(\boldsymbol{Z}(t), \boldsymbol{V}(t), t). \tag{1.1}$$

We have written $\partial_r L$, $r \in \{1, \ldots, 7\}$, for the partial derivative with respect to the r argument (or r slot) of the function L. The partial derivative with respect to the time variable is $\partial_7 L$, although we shall usually use the alternative notation $\partial_t L$. Omitting the time variable gives a neater version of (1.1):

$$F^i = \partial_i L(\boldsymbol{Z}, \boldsymbol{V}, \,.), \qquad P^i = \partial_{i+3} L(\boldsymbol{Z}, \boldsymbol{V}, \,.), \tag{1.2}$$

where the dots in the seventh slots mark the missing variable t (cf. the remarks after I(2.4)). Eqs.(1.2) make sense because both the left-hand and right-hand sides are functions of the single variable t. The common notations $F^i = \partial L/\partial Z^i$ and $P^i = \partial L/\partial V^i$ make no sense at all except as shorthand forms of (1.2).

Mathematical note (i) If its partial derivatives are to be defined, L must be defined and differentiable on an open set of $\mathbb{R}^7$. We usually assume that it is C^2 there (i.e. all its second partial derivatives are defined and continuous), but less stringent conditions are sometimes possible, and necessary.

Mathematical note (ii) In (1.1) and (1.2), one must *first* perform the partial differentiations, and *then* substitute the $\boldsymbol{Z}$ and $\boldsymbol{V}$ that correspond to the path of the particle. To be mathematically correct, one should use different symbols for the variables in $\mathbb{R}^7$ and those on the path (one might write $L(\boldsymbol{x}, \boldsymbol{U}, s)$, for example).

The Lagrangian equations of motion of a particle have the same form as the Newtonian equations: $D\boldsymbol{P} = \boldsymbol{F}$, where D is the time derivative. If we define $E = \sum_i V^i P^i - L(\boldsymbol{Z}, \boldsymbol{V}, \,.)$ (a function of time only), we have (dropping a few arguments to save writing)

$$\begin{aligned} DE &= \sum_i \left[(DV^i)P^i + V^i DP^i - V^i \partial_i L - (DV^i)\partial_{i+3} L\right] - \partial_t L \\ &= \sum_i \left[(DV^i)P^i + V^i F^i - V^i F^i - (DV^i)P^i\right] - \partial_t L = -\partial_t L, \end{aligned}$$

or (putting the arguments back)

$$DE(t) = -\partial_t L(\boldsymbol{Z}(t), \boldsymbol{V}(t), t). \tag{1.3}$$

If $\partial_t L = \partial_7 L = 0$, one says that L does not depend *explicitly* on time. In this case, $DE = 0$ and E is a constant of the motion – it has the same value at all points of the path of the particle.

If you have not met Lagrangian mechanics before, it may seem rather difficult and abstract. Do not worry! A virtue of the Lagrangian method is that it allows you to do mechanics in a mechanical way – almost without thinking. The following example will show how easy it is.

For a free particle in special relativity, the Lagrangian is $L(\boldsymbol{Z}, \boldsymbol{V}, t) = -mc^2/\gamma$, where m, c, and γ are defined as before. Since L is in fact independent of $\boldsymbol{Z}$ and t, og $F^i = 0$ from (1.2), and

$$P^i = m\gamma V^i, \qquad DP^i = 0. \tag{1.4}$$

The canonical momentum is therefore the same as the mechanical momentum previously defined, and is a constant of the motion. From (1.3) one shows that

$$E = m\gamma c^2 \tag{1.5}$$

is a constant of the motion. As before, one calls E the *energy* of the particle. When the particle is at rest, one has $\gamma = 1$ and $E = mc^2$ is the *rest energy* of the particle.

2. Lagrangian mechanics in a gravitational field

We are now going to find the proper mechanics (that is, the correct Lagrangian) for a particle in a gravitational potential. We adopt, as always, the conservative principle that traditional ideas are to be altered as little as possible.

The special relativistic Lagrangian for a free particle is $L(\boldsymbol{Z}, \boldsymbol{V}, t) = -mc^2(1 - |\boldsymbol{V}|^2 c^{-2})^{1/2}$. It is also, with a small change of interpretation, the Lagrangian for a particle in a gravitational potential. We have only to replace m and c by the Newtonian quantities m_N and c_N, as given by II(7.7) and II(7.8). There is no need to change $|\boldsymbol{V}|^2$ because the components of $\boldsymbol{V}$ are already Newtonian: one has $V^r = DZ^r$, which is the derivative of a space coordinate with respect to the time coordinate. Writing $c_E = c$ as before, and $\phi - \phi_0 = \psi$, og

$$\begin{aligned} L(\boldsymbol{Z}, \boldsymbol{V}, t) = -m_E c^2 \exp\Big[(2\tau - 2\lambda - \mu)\psi(\boldsymbol{Z}, t)\Big] \\ \Big\{1 - |\boldsymbol{V}|^2 c^{-2} \exp\Big[(2\lambda - 2\tau)\psi(\boldsymbol{Z}, t)\Big]\Big\}^{1/2}. \end{aligned} \tag{2.1}$$

The mass m_E, which is the experimentally measurable, proper mass of the particle, is assumed to be constant. The canonical momentum $\boldsymbol{P}$ is found from (1.2):

$$\boldsymbol{P} = m_N \gamma_N \boldsymbol{V}, \qquad \gamma_N = (1 - |\boldsymbol{V}|^2 c_N^{-2})^{-1/2}, \tag{2.2}$$

which is exactly similar to the special relativistic expression (1.4). The force acting on the particle – which we denote by $\boldsymbol{W}$ and call the *gravitational force* – is also found from (1.2):

$$\boldsymbol{W} = -m_N c_N^2 \gamma_N \big[2\tau - 2\lambda - \mu + (\tau - \lambda - 1)|\boldsymbol{V}|^2 c_N^{-2}\big] \nabla\psi(\boldsymbol{Z},\ .). \tag{2.3}$$

The equation of motion of the particle is $D\boldsymbol{P} = \boldsymbol{W}$. If an additional, nongravitational force $\boldsymbol{N}$ acts on the particle, then the equation of motion becomes

$$D\boldsymbol{P} = \boldsymbol{W} + \boldsymbol{N}. \tag{2.4}$$

However, in this chapter we are concerned with gravitational forces only, and $\boldsymbol{N} = 0$.

The Lagrangian (2.1) does not depend explicitly on t, so (1.3) implies that $E = \sum_i V^i P^i - L(\boldsymbol{Z}, \boldsymbol{V},\ .)$ is a constant of the motion. Og

$$E = m_N \gamma_N c_N^2, \tag{2.5}$$

which is of exactly the same form as the special relativistic expression (1.5). Eqs. (2.2) and (2.5) give

$$E^2 - |\boldsymbol{P}|^2 c_N^2 = m_N^2 c_N^4, \tag{2.6}$$

which is again exactly similar to a special relativistic equation.

In Section II.7, we made the convention that any universal constant written without a subscript E or N has its empirical value, and that any other quantity written without such a subscript has its Newtonian value: that $c = c_E$, $m = m_N$, $\gamma = \gamma_N$, etc. This saves us a good deal of writing, although for clarity we shall sometimes – as in (2.2)–(2.6) – put in redundant E's and N's. As before, Newtonian quantities are defined with respect to a ϕ_0 frame: they are the same as empirical quantities in a region where $\phi \approx \phi_0$ (provided that they are local, insofar as one can make a correspondence between empirical and mathematical quantities, etc.).

To determine the gravitational potential, we need a field equation. Being good conservatives, we adopt the Poisson equation I(2.8), which we interpret as holding in a ϕ_0 frame. A few small changes are required. We replace the Newtonian gravitational potential Φ by the dimensionless potential $\phi = \Phi c^{-2}$ (or equivalently by $\psi = \phi - \phi_0$). In addition, we replace the mass density ρ on the right-hand side of the equation by the energy density ϵ divided by the square of the speed of light. This is more general, because it allows the energy of a field (e.g. an electromagnetic field) to be a source of the gravitational field. It also seems more physically reasonable, because energy is a conserved quantity (at least in the special relativistic limit) while mass is not. For sources that consist only of stationary masses, the generalization makes no difference, since $\epsilon_E = \rho_E c^2$ by the Einstein mass-energy equivalence.

To summarize a whole paragraph by a short equation, the field equation is

$$\nabla^2 \phi = 4\pi k_N \epsilon, \tag{2.7}$$

where ϵ is the energy density of the sources in ϕ_0 units (i.e. in the Newtonian units that correspond to a ϕ_0 frame), and $k = Gc^{-4}$ is a universal constant. The left-hand side of (2.7) is in ϕ_0 units (because ϕ is dimensionless), and we must therefore write k_N on the right-hand side. The dimensions of k are $[k] = [\mathcal{L}^{-1}\mathcal{T}^2\mathcal{M}^{-1}]$, and II(7.2) and II(7.6) give

$$k_N = k \exp\Big[(\lambda - 2\tau + \mu)(\phi - \phi_0)\Big]. \tag{2.8}$$

The field equation (2.7) does not involve time derivatives: we are still dealing with gravitostatics. We shall however stretch the theory, just as one stretches the Newtonian theory, to apply to quasistatic situations.

3. Determination of τ, λ, and μ

To determine completely the relationship between empirical and Newtonian quantities, we must find the constants τ, λ, and μ in II(7.3)–II(7.5). We could just write down their values, and then show that the theory gives results in agreement with experiment, but it is more satisfying to derive them from physical arguments.

Newtonian limit

The theory must reduce to the Newtonian theory when V/c and ϕ are small (their magnitudes much less than 1). This means that the Lagrangian (2.1) must be equivalent to the Newtonian Lagrangian $(1/2)m_E|\boldsymbol{V}|^2 - m_E c^2 \phi(\boldsymbol{Z}, \,.)$. If we expand (2.1) in powers of V/c_E and ϕ and neglect small terms, we find that $L(\boldsymbol{Z}, \boldsymbol{V}, t) \approx (1/2)m_E|\boldsymbol{V}|^2 - m_E c^2 - m_E c^2(2\tau - 2\lambda - \mu)\psi(\boldsymbol{Z}, t)$. This agrees with the Newtonian Lagrangian, apart from insignificant constants, provided that

$$2\tau - 2\lambda - \mu = 1. \tag{3.1}$$

Active gravitational mass

If the sources of the gravitational field have spatially bounded support (that is, if ϵ vanishes outside some finite spatial region), then (2.8) and the Poisson equation

(2.7) imply that

$$\psi(\boldsymbol{x}) - \chi_0 = k\mathcal{E}r^{-1} + O(r^{-2}), \tag{3.2}$$

as $r = |\boldsymbol{x}| \to \infty$, where χ_0 is a constant and

$$\mathcal{E} = \int \epsilon(\boldsymbol{x}) \exp\Big[(\lambda - 2\tau + \mu)(\phi(\boldsymbol{x}) - \phi_0)\Big]\, d^3x. \tag{3.3}$$

It is usual to call $\mathcal{E}c^{-2}$ the *active gravitational mass* of the sources, although from our point of view it would be more reasonable to call $\mathcal{E}$ the *active gravitational energy*. (Note that one cannot define the active gravitational mass to be $\mathcal{E}c_N^{-2}$ because $\mathcal{E}$ is not a local quantity, and consequently c_N is not well defined.) The active gravitational mass is a measure of the strength of the gravitational field at great distances from the sources.

Some explanation may be called for. Eq.(3.2) is the *monopole approximation* to the potential, which you have probably met in electrostatics. We say that a function f of r is of order r^n as $r \to \infty$ (and we write $f(r) = O(r^n)$ as $r \to \infty$) if $f(r)/r^n$ is bounded as $r \to \infty$). This means that there are constants a and b such that $| f(r)/r^n | < b$ for all $r > a$. In the integral (3.3), d^3x stands for $dx^1dx^2dx^3$, and the integration is over all values of x^1, x^2, and x^3.

We now make the important assumption that the active gravitational mass of a source is proportional to its total ϕ_0 energy (and the constant of proportionality is the same for all sources). As usual, there is nothing really new here – the assumption is implicit in Newtonian theory. But because of our more general definition of the active gravitational mass, it does have wider consequences.

The total ϕ_0 energy of a source is $\int \epsilon(\boldsymbol{x})\, d^3x$, where again the integration is over all values of x^1, x^2, and x^3. If this is to be proportional to (3.3) for an arbitrary, bounded source, we must have

$$\lambda - 2\tau + \mu = 0. \tag{3.4}$$

Eqs.(3.1) and (3.4) imply that $\lambda = -1$, and $\mu - 2\tau = 1$.

Planck's Law

We have shown that the Newtonian energy E of a particle in a static gravitational potential is a constant of the motion (it has the same value everywhere on the path). Similarly, the Newtonian frequency ν of an electromagnetic wave is constant along its path. We assume, as in primitive quantum theory, that a photon is a particle of energy E which is associated with a wave of frequency ν, and that *Planck's Law* $E = h_N\nu$ is valid. (We have to write h_N because E and ν are Newtonian quantities.) Since E and ν are constant along the path, so is h_N. The dimensions of h are $[h] = [\mathcal{M}\mathcal{L}^2\mathcal{T}^{-1}]$, and II(7.2) and II(7.6) imply that in a ϕ_0

frame $h_N = h \exp[(-2\lambda + \tau - \mu)(\phi(\boldsymbol{x}) - \phi_0)]$. The empirical value of the universal constant h is of course the same everywhere, and it follows that

$$2\lambda - \tau + \mu = 0. \tag{3.5}$$

Eqs.(3.5) and (3.1) imply that $\tau = 1$: which is the value required to explain the measured value of the gravitational red shift. Using (3.4), og

$$\tau = 1, \qquad \lambda = -1, \qquad \mu = 3. \tag{3.6}$$

In the argument that led to (3.5), we assumed that the energy of a photon is a constant of the motion, like that of a material particle. A photon is not, of course, a material particle, but we can perhaps regard it as a limiting case of a particle of mass m and speed $V = |\boldsymbol{V}|$. We consider a sequence of particles for which $m \to 0$ and $V \to c_N$ in such a manner that $E = m\gamma c_N^2$ remains constant. The photon is regarded as being the limit of the sequence, with $m = 0$, $V = c_N$, and E a constant of the motion. The momentum $\boldsymbol{P}$ of the photon is parallel to its velocity, and has magnitude E/c_N, from (2.6). Although our "photon" has no spin, it may serve as an adequate, classical model. (If all this seems too mathematical, one can simply regard a photon as being a particle of very small, but finite mass.)

4. Summary: Newtonian and empirical quantities

Now that we know the values (3.6) of the constants τ, λ, μ, we can write the relationships between Newtonian and empirical quantities in an explicit form. As usual, the Newtonian quantities are defined with respect to a ϕ_0 frame so that, for any local quantity q, one has $q_N = q_E$ in a region where $\phi \approx \phi_0$. We shall again write $\psi = \phi - \phi_0$. In what follows, we always consider spatially bounded sources, and it is convenient to choose the frame so that $\phi \to \phi_0$ and $\psi \to 0$ at great distances from the source (i.e. we take $\chi_0 = \phi_0$ in (3.2)).

Time, length, mass

For quantities with the dimensions of time, length, and mass, eqs.II(7.3)–II(7.5) become

$$\mathrm{t} = \mathrm{t}_E e^{-\psi}, \qquad l = l_E e^{\psi}, \qquad m = m_E e^{-3\psi}. \tag{4.1}$$

Speed of light

For c, the speed of light in vacuum, one has $[c] = [\mathcal{L}\mathcal{T}^{-1}]$, and II(7.8) becomes

$$c_N = c e^{2\psi}, \tag{4.2}$$

where $c = c_E$ because c is a universal constant. We define $c/c_N = e^{-2\psi}$ to be the *refractive index of space*. It tells how quickly electromagnetic waves, or light signals, travel in a ϕ_0 frame. The solar system, for example, is like a crystal ball whose refractive index decreases as one goes outwards from the Sun, approaching the value 1 at very great distances.

Eq.(4.2) has measurable consequences in radar time-delay experiments. Radar signals from Earth are bounced off other planets, and the time taken for the round trip Earth-planet-Earth is measured. The results agree with the calculated motions of the planets if the speed of the signals is assumed to be c_N as calculated from (4.2).

We are not of course asserting that the *measured* speed of the radar pulses is c_N. The paths of the pulses, and of the planets, are calculated in a Newtonian frame, and in that frame the speed of the pulses is c_N.

Exercise 4.1 A radar pulse from Earth is reflected from Mercury and returns to Earth. The radius vector from the Sun to Earth makes an angle of 120° with that from the Sun to Mercury. Calculate the round-trip time of the radar pulse (a) assuming that its speed in the Newtonian frame is c, (b) assuming that it is c_N as given by (4.2). You may assume that the orbits of Earth and Mercury are coplanar circles centred at the Sun, of radius 1.50×10^8 km and 5.8×10^7 km, respectively. ∎

Action, angular momentum

As we showed in the derivation of (3.5), Planck's constant is the same in empirical and Newtonian units, $h = h_N$. More generally, action and angular momentum, which have the same dimensions $[\mathcal{M}\mathcal{L}^2\mathcal{T}^{-1}]$, are the same in empirical and Newtonian units.

Gravitational constant

It follows from (2.8) and (3.4) that the constant $k = Gc^{-4}$ which appears in the field equation (2.7) is the same in empirical and Newtonian units, $k = k_N$. Its value is $k = 8.261 \times 10^{-45}\text{kg}^{-1}\text{m}^{-1}\text{s}^2$. We shall call k the *gravitational constant*; we shall continue to call G the *Newtonian gravitational constant.*

Energy, Lagrangian, momentum

The dimensions of energy are $[E] = [\mathcal{M}\mathcal{L}^2\mathcal{T}^{-2}]$, and Newtonian and empirical energies are related by

$$E = E_E e^{\psi}, \tag{4.3}$$

which has the same exponent as the length equation, from (4.1). Note that (4.3) holds (like all such equations) only for *local* quantities: it is not valid for the energy found by integrating an energy density over a region where ϕ varies, for example.

The particle Lagrangian (2.1), which has the dimensions of energy, becomes

$$L(\boldsymbol{Z}, \boldsymbol{V}, t) = -m_E c^2 e^{\psi(\boldsymbol{Z},t)} \Big[1 - |\boldsymbol{V}|^2 c^{-2} e^{-4\psi(\boldsymbol{Z},t)}\Big]^{1/2}. \tag{4.4}$$

As in (2.1), we have allowed ψ to depend on t.

The momentum (2.2) of a particle becomes (as usual, we write m for m_N and γ for γ_N)

$$\boldsymbol{P} = m\gamma \boldsymbol{V} = m_E e^{-3\psi(\boldsymbol{Z},.)} \Big[1 - |\boldsymbol{V}|^2 c^{-2} e^{-4\psi(\boldsymbol{Z},.)}\Big]^{-1/2} \boldsymbol{V}, \tag{4.5}$$

where $\boldsymbol{P}$ and $\boldsymbol{Z}$ are functions of t, and the dot in $(\boldsymbol{Z}, .)$ denotes the missing variable t. In doing calculations, one can often omit the arguments and write ψ instead of $\psi(\boldsymbol{Z}, .)$, etc., but one must put them in whenever confusion threatens.

Force

Force, which has the dimension $[\mathcal{MLT}^{-2}]$, is the same in empirical and Newtonian units (since $\mu + \lambda - 2\tau = 0$ from (3.6)). This has the fortunate consequence that one can do statics without worrying whether the forces are in empirical or Newtonian units. However, the *moments* of the forces are not the same.

Exercise 4.2 A horizontal force of 100Nt has a torque of 1000Nt m about a point P in the same horizontal plane. Calculate the change in the torque if the Newtonian gravitational potential is increased by $100\text{m}^2\text{sec}^{-2}$ (e.g. by lifting the entire system through about 10m). Is this change in torque measurable? ■

The gravitational force (2.3) on a particle becomes

$$\boldsymbol{W} = -E[1 + |\boldsymbol{V}|^2 c_N^{-2}]\nabla\psi(\boldsymbol{Z}, .), \tag{4.6}$$

where $E = m_N c_N^2 \gamma_N = m c_N^2 \gamma$ from (2.5). If ψ does not depend on t, and if no non-gravitational forces act on the particle, then E is a constant of the motion (the *energy* of the particle).

For a slow moving particle, (4.6) reduces to

$$\boldsymbol{W} = -E\nabla\psi(\boldsymbol{Z}, .) = -Ec^{-2}\nabla\Phi(\boldsymbol{Z}, .), \tag{4.7}$$

which is the same as in the Newtonian theory. For a particle moving at or near the speed of light, (4.6) becomes $\boldsymbol{W} = -2E\nabla\psi(\boldsymbol{Z}, .)$ – which is twice as much as in the Newtonian theory. In other words, the gravitational force on a photon is twice the force on a slowly moving particle of the same energy.

5. Bending of light

We already know how to calculate the bending (or deflection) of an electromagnetic wave – a light beam or a radar pulse – in a gravitational field. The refractive index of space in a Newtonian frame is $c/c_N = e^{-2\psi}$, from (4.2), and we can use the standard methods of ray optics provided that the wavelength of the wave is small compared with distances over which ψ changes appreciably.

Exercise 5.1 Calculate the deflection of starlight by the Sun using the techniques of ray optics. [You can appeal to Fermat's principle or Snell's law. In the latter case, regard the space around the Sun as a set of concentric, spherical shells, each of constant refractive index. Your result should agree with (5.3).] ∎

By using the previous, primitive model of a photon, we can also calculate the deflection by the methods of particle mechanics. We regard a photon as a limiting case of a material particle (or as a particle of extremely small mass). Its energy E is a constant of the motion; its momentum $\boldsymbol{P}$ is parallel to the velocity, and has magnitude $P = |\boldsymbol{P}| = E/c_N$, from (2.6). The gravitational force on the photon is $\boldsymbol{W} = -2E\nabla\psi(\boldsymbol{Z},\ .)$, from (4.6), and the equation of motion is $D\boldsymbol{P} = \boldsymbol{W}$.

To calculate the deflection of a photon as it passes the Sun, we assume that the gravitational potential is, to a sufficient approximation, static and spherically symmetric. From (3.2), og

$$\psi(\boldsymbol{x}) = \phi(\boldsymbol{x}) - \phi_0 = -k\mathcal{E}r^{-1}, \tag{5.1}$$

where $r = |\boldsymbol{x}|$, the origin of spatial coordinates is taken at the centre of the Sun, k is the gravitational constant, and $\mathcal{E} = \int \epsilon(\boldsymbol{x})\, d^3x$ is the total Newtonian energy of the Sun.

We calculate the deflection of a photon whose path begins and ends at a great distance from the Sun. We find the change $\Delta\boldsymbol{P}$ in the photon's momentum by integrating the equation of motion $D\boldsymbol{P} = \boldsymbol{W}$ along the path. Substituting the expression for $\boldsymbol{W}$, og

$$\Delta\boldsymbol{P} = -2E\int \nabla\psi(\boldsymbol{Z}(t))\, dt, \tag{5.2}$$

where the integation may be taken from $-\infty$ to ∞, and ψ is given by (5.1). Since ψ and $\nabla\psi$ are everywhere very small, the path of the photon is almost a straight line, and its speed is always very close to c. We can therefore evaluate (5.2) to a good approximation by assuming that the photon moves in a straight line at constant speed. By choosing the ϕ_0 frame suitably (rotate the spatial axes and change the zero of time if necessary), og $\boldsymbol{Z}(t) = (ct,\ b,\ 0)$, where the constant b is the ϕ_0 distance of closest approach of the photon to the origin (i.e. to the centre of the Sun). Differentiating (5.1), og $\nabla\psi(\boldsymbol{Z},t) = k\mathcal{E}(c^2t^2+b^2)^{-3/2}(ct,b,0)$. Evaluating the integral in (5.2), og $\Delta P^1 = \Delta P^3 = 0$, and $\Delta P^2 = -4Ek\mathcal{E}/bc$. The photon is therefore deflected towards the Sun through an angle

$$|\Delta\boldsymbol{P}|/P = 4\mathcal{E}k/b = 4GM/bc^2, \tag{5.3}$$

where M is the mass of the Sun and we have made the approximation that $\mathcal{E} = Mc^2$.

Exercise 5.2 Complete the above proof, and so verify (5.3). [Note that $\int_{-\infty}^{\infty}(1+z^2)^{-3/2}\,dz = 2$.] ■

Exercise 5.3 Calculate the deflection of starlight from the Sun as observed from the orbit of the Earth. [Make a suitable change in the limits of integration, and estimate the correction to (5.3). Do not consider the problem of measuring the deflection from the moving Earth.] ■

6. Perihelion advance

We can find the orbit of a particle in the potential (5.1) by solving the equations of motion $D\boldsymbol{P} = \boldsymbol{W}$, with $\boldsymbol{W}$ given by (4.6). This is the analog of the Kepler problem in the Newtonian theory. One would expect that when ψ and $\nabla\psi$ are small, as in the solar system, the bounded orbits should be slightly perturbed ellipses.

Using (4.5) for the momentum of the particle, we write the equations of motion as

$$\begin{aligned} D\boldsymbol{P} &= D\Big[m_E e^{-3\psi(\boldsymbol{Z},\,\cdot)}\gamma\boldsymbol{V}\Big] = -E\Big[1+|\boldsymbol{V}|^2c_N^{-2}\Big]\nabla\psi(\boldsymbol{Z},t) \\ &= -E\Big[1+|\boldsymbol{V}|^2c^{-2}e^{-4\psi(\boldsymbol{Z},\,\cdot)}\Big]k\mathcal{E}|\boldsymbol{Z}|^{-3/2}\boldsymbol{Z}. \end{aligned} \tag{6.1}$$

This looks forbidding, but we solve it by the usual trick of looking for conserved quantities (i.e. constants of the motion). Of course, we already know from (2.5) and (4.3) that $E = m_Ec^2e^{\psi}\gamma$ is conserved.

By taking the vector product of (6.1) with $\boldsymbol{Z}$, and noting that $\boldsymbol{V}\times P = 0$, we show that $D(\boldsymbol{Z}\times P) = 0$. Hence $\boldsymbol{j} = \boldsymbol{Z}\times P$ is a constant of the motion, as in Newtonian theory. We call $\boldsymbol{j}$ the ***(orbital) angular momentum*** of the particle. We assume that $j = |\boldsymbol{j}| \neq 0$ (the orbit does not pass through the origin!); the vector $\boldsymbol{j}$ is then normal to the plane of the orbit. We introduce polar coordinates r and θ in this plane, and find that $|\boldsymbol{V}|^2 = r^2(D\theta)^2 + (Dr)^2$ and $j = m\gamma r^2D\theta$. We eliminate time derivatives by regarding r as a function of θ, so that $(dr/d\theta)^2 = (Dr)^2(D\theta)^{-2} = |\boldsymbol{V}|^2(D\theta)^{-2} - r^2$. Writing $|\boldsymbol{V}|^2 = c_N^2(1-\gamma^{-2})$, $D\theta = j/m\gamma r^2$, we get

$$(dr/d\theta)^2 = c_N^2(1-\gamma^{-2})m^2\gamma^2r^4/j^2 - r^2.$$

Finally, we eliminate γ by using $\gamma = E/mc_N^2$, and define $u = 1/r$ so that $(du/d\theta)^2 = r^{-4}(dr/d\theta)^2$. This gives

$$(du/d\theta)^2 = j^{-2}(E^2c_N^{-2} - m^2c_N^2) - u^2.$$

We have $\psi = -k\mathcal{E}u$ from (5.1). We use (4.1) and (4.2), expand in powers of u and neglect terms of third degree, and find that

$$(du/d\theta)^2 = (A-1)u^2 + 2Bu + C, \tag{6.2}$$

where

$$A = (8E^2 - 2m_E^2c^4)k^2\mathcal{E}^2j^{-2}c^{-2}, \qquad B = (2E^2 - m_E^2c^4)k\mathcal{E}j^{-2}c^{-2},$$
$$C = (E^2 - m_E^2c^4)j^{-2}c^{-2}. \tag{6.3}$$

Because we are assuming that ψ is everywhere small, we may take $k\mathcal{E} \approx \mathrm{GMc}^{-2}$, where M is the total mass (Newtonian or empirical) of the source of the potential. If in addition Vc_N^{-1} is everywhere small, we may write $E \approx m_Ec^2 \approx mc^2$. We then have

$$A \approx 6(GMmj^{-1}c^{-1})^2, \qquad B \approx \mathrm{GMm}^2j^{-2}, \qquad C \approx 0. \tag{6.4}$$

Differentiating (6.2) with respect to θ, og $d^2u/d\theta^2 = (A-1)u + B$. If $|A| < 1$, the solution is

$$u = B(1-A)^{-1}\{1 - \beta\cos[(1-A)^{1/2}\theta + \delta]\}, \tag{6.5}$$

where β and δ are constants, and one may take $\beta > 0$. The orbits described by (6.5) are bounded if $0 < \beta < 1$ (that is, $r = 1/u$ is everywhere finite); the orbits are unbounded if $\beta \geq 1$.

If $0 < \beta < 1$, the maxima of u (and the minima of $r = u^{-1}$) occur when $\theta = \theta_n$, where $(1-A)^{1/2}\theta_n + \delta = (2n+1)\pi$, for $n \in \{0, \pm1, \pm2, \ldots\}$. Hence $\theta_{n+1} - \theta_n = 2\pi(1-A)^{-1/2} \approx 2\pi + \pi A$ if $|A| \ll 1$. From (6.4) one has

$$\pi A \approx 6\pi(GMmj^{-1}c^{-1})^2. \tag{6.6}$$

One usually calls πA the *relativistic perihelion advance* (in radians per revolution). For a planet of the Sun, a point on the orbit where r is a minimum is called *a perihelion point*, or simply *perihelion* (from the Greek meaning *near the Sun*). For an Earth satellite, one may use the term *perigee*, and for a planet of a star *periastron*, etc. – but most people use *perihelion* for everything.

Exercise 6.1 Fill in the gaps in the above proofs, and so verify eqs.(6.1)–(6.6). ■

The motion of planets in the solar system is complicated because they all interact with one another. It required two centuries of effort before the predictions of the Newtonian theory were confidently established, and shown to agree almost perfectly with observation. Almost, but not quite – there are a few, very small anomalies, of which the most significant is the perihelion advance of Mercury. This residual discrepancy is accounted for quite well by (6.6) (see also Section IV.5).

Our calculation of the relativistic perihelion advance is typical of the methods of *perturbation theory*. We calculate the motion of a particle in the potential (5.1), and compare it with the Newtonian motion in the same potential (where there is no perihelion advance). We then argue that the difference between the motions is the same as if we had dealt with the more complex system where all the planets interact and the Sun moves with respect to the centre of mass. The slogan is *small perturbations act independently of one another*. It is almost certainly true in the present case, but a proper mathematical justification is difficult, and will not be attempted here [Taff 85]. A brief account of perturbation theory is given in Appendix F.

Exercise 6.2 Calculate the relativistic perihelion advance for Mercury. Give your result in radians per mercurial year,[2] and in minutes of arc per century. Also calculate it in metres per orbit (the distance between successive perihelion points). [The semimajor axis of Mercury's orbit is 5.8×10^{10}m, its eccentricity is 0.206, and the orbital period is 88 days. As a first approximation, if the reader has forgotten all about eccentricity and the Kepler problem, he can assume that the orbit is almost circular.] ∎

7. Historical note

Three of the effects that we have discussed – the gravitational redshift, the deflection of light, and the perihelion advance – are often called the *classical tests* of relativistic gravity because of their long history. Although it is much more recent, we shall regard the radar time delay as another classical test because it is so closely related to the deflection of light. The earliest of the effects to be established was the anomalous perihelion advance, calculated by Leverrier in 1845. The redshift was proposed by Einstein in 1907, but proved to be extremely difficult to measure. Many attempts were made to measure the spectral lines of the Sun and of the white dwarfs Sirius B and 40 Eridani B, but the results were inconclusive until the 1960's [Bertotti 1962]; laboratory measurements of the effect date from the same period. The light deflection was first measured, with about 10 percent accuracy, at the solar eclipse of 29 May 1919.

The reader can now judge the correctness of the earlier assertion that all the classical tests of relativistic gravity can be easily calculated from a theory that is almost identical with the Newtonian. The only significant, additional assumptions are that the theory must reduce to special relativity in the absence of a gravitational field, and that physical quantities may depend on the gravitational potential; the first is obvious, and the second is forced by the existence of the gravitational redshift. We call this slightly modified Newtonian theory *gravitostatics*.

If intelligent aliens had visited Earth around 1910 and assessed the state of gravitational theory, they would no doubt have predicted that gravitostatics as we have described it would soon be discovered, to be followed by more complete theories of gravity. In the event, this did not happen. Gravitostatics remained undiscovered for half a century [Rastall 60, 79], and a much more difficult, general, and speculative theory was invented by Einstein.[3] It is interesting to read the histories of those times, and wonder how it happened. It was perhaps partly that special relativity was very new, with its emphasis on the equivalence of inertial charts with arbitrary relative velocity. Enthusiastic young theorists were in no mood to restrict themselves to static gravitational fields in a preferred frame of reference.

It may seem perverse to describe the Einstein theory, which is the most widely accepted theory of gravity, as speculative. I do not intend to be derogatory – it is in fact a brilliant, speculative theory. However, one should contrast it with the very conservative kind of theory that we have been developing, in which new assumptions are kept to a minimum. We cannot discuss the mathematical aspects of the Einstein theory here[4] (there is a brief account in Chapter IX). Its main weakness, as with all relativistic theories of gravity, is that there is hardly any empirical evidence for it. We have seen that the classical tests are explained by gravitostatics, and almost all the more recent experiments are explained by an extension of gravitostatics (the postnewtonian theory) which is developed in later chapters. Apart from agreement with the postnewtonian theory, the only good evidence for the Einstein theory at present is from measurements of binary pulsars.

We should not conclude that relativistic theories are unworthy of study – they are necessary for any complete description of gravity, and in particular of gravitational waves and of the motion of very massive bodies – but at present there is little reason to believe any of them. However, the experimental evidence may improve dramatically within a few years.

8. Geodesic motion

We are going to show that the motion of a particle in a gravitational field can be described geometrically in terms of the spacetime metric. We shall introduce the covariant derivative, which plays an important role in relativistic theories of gravity and in differential geometry. *This section depends on Section II.8. Readers who omitted Section II.8 should also omit this one, at least until they reach Chapter V.*

The equation of motion of a particle on which only gravitational forces act is found from the Lagrangian (4.4). One has $\boldsymbol{DP} = \boldsymbol{W}$, where the momentum $\boldsymbol{P}$ and the gravitational force $\boldsymbol{W}$ are given by (4.5) and (4.6), and hence

$$D(m_E e^{-3\psi(\boldsymbol{Z},\,.)}\gamma \boldsymbol{V}) = -m_E c^2 e^{\psi(\boldsymbol{Z},\,.)}\gamma\Big[1+|\boldsymbol{V}|^2 c_N^{-2}\Big]\nabla\psi(\boldsymbol{Z},\,.). \tag{8.1}$$

As before, m_E is constant, $\gamma = \Big[1 - |\boldsymbol{V}|^2 c^{-2} e^{-4\psi(\boldsymbol{Z},\,.)}\Big]^{-1/2}$, and $\boldsymbol{Z}$ and $\boldsymbol{V}$ are functions of the time coordinate t.

Another way of representing the path of the particle is as a curve $z : \mathbb{I} \to \mathbb{R}^4, \mathbb{I} \subset \mathbb{R}$. We write $z(u) = (z^0(u), z^1(u), z^2(u), z^3(u))$, where u is an invariant parameter (the same in all charts), $z^m(u)$ are the spatial coordinates of the particle, and $z^0(u)/c$ is its time coordinate. We note that the notation here is different from that in Chapters I and II, where $z^m(P)$ is the m coordinate of the spatial point P.

The equations of motion are derived from the invariant Lagrangian

$$\mathcal{L}(z, Dz,\ .) = \sum_{\mu,\nu} g_{\mu\nu}(z) Dz^\mu Dz^\nu. \tag{8.2}$$

(As usual, lower-case Greek indices are to be summed over their range $\{0,1,2,3\}$, and lower-case Latin over $\{1,2,3\}$.) The equations of motion are $D\partial_{\mu+4}\mathcal{L}(z, Dz,\ .) = \partial_\mu \mathcal{L}(z, Dz,\ .)$, or

$$\sum_\nu 2D(g_{\mu\nu}(z)v^\nu) = \sum_{\pi,\rho} \partial_\mu g_{\pi\rho}(z) v^\pi v^\rho, \tag{8.3}$$

where $v^\mu = Dz^\mu = dz^\mu/\mathrm{du}$. (One can of course regard (8.3) as the Euler equations of a variational principle with integrand (8.2).) Note that (8.3) is invariant under the transformation $u \mapsto au + b$, where a and b are constants, $a \neq 0$.

One finds a first integral of (8.3) by multiplying by v^μ and noting that $\sum_\mu v^\mu \partial_\mu g_{\pi\rho}(z) = Dg_{\pi\rho}(z)$. Og $\sum_{\mu,\nu} Dg_{\mu\nu}(z)v^\mu v^\nu + \sum_{\mu,\nu} 2g_{\mu\nu}(z)v^\mu Dv^\nu = 0$, and $\sum_{\mu,\nu} D(g_{\mu\nu}(z)v^\mu v^\nu) = 0$, and hence $\sum_{\mu,\nu} g_{\mu\nu}(z)v^\mu v^\nu$ is a constant of the motion. For a material particle, the constant is negative, and by choosing u appropriately og

$$\sum_{\mu,\nu} g_{\mu\nu}(z)v^\mu v^\nu = -c^2. \tag{8.4}$$

In this case we write $u = \tau$, and call τ the *proper time* (do not confuse with the parameter τ of (3.6)!). When $u = \tau$, the vector field v with components v^μ is the *4 velocity* of the particle.

The 3 velocity is given by $\boldsymbol{V} = D\boldsymbol{Z}$, where $\boldsymbol{Z}(t) = \boldsymbol{z}(\tau)$. Since $v^0 = c\,dt/d\tau$,

$v^m(\tau) = (dt/d\tau)V^m(t) = v^0(\tau)V^m(t)c^{-1}$, eq.(8.4) implies that

$$v^0c^{-1} = \Big\{-g_{00}(z) - \sum_m 2g_{m0}(z)V^mc^{-1} - \sum_{m,n} g_{mn}(z)V^mV^nc^{-2}\Big\}^{-1/2}, \tag{8.5}$$

where we have omitted arguments τ and t and chosen the positive sign of the square root.

In our case, the $g_{\mu\nu}$ are given by II(8.2) and II(8.7), and (8.3) with $\mu = m$ becomes

$$2D(e^{-2\psi(z)}v^m) = \partial_m e^{-2\psi(z)} \sum_p v^pv^p - \partial_m e^{2\psi(z)}(v^0)^2. \tag{8.6}$$

Eq.(8.5) becomes $v^0(\tau)/c = e^{-\psi(z)}\gamma$, and it follows that $v^m(\tau) = V^m(t)e^{-\psi(z)}\gamma$. Using the fact that m_E is constant, one shows that (8.6) is equivalent to (8.1).

Exercise 8.1 Derive (8.1) from (8.6). ■

We define the functions $g^{\rho\mu}$ to be the solutions of the equations $\sum_\mu g^{\rho\mu}g_{\mu\nu} = \delta_{\rho\nu}$ (i.e. they are the elements of the matrix inverse to the matrix with elements $g_{\mu\nu}$). To elucidate the geometrical significance of (8.3), we first write $\sum_\nu D(g_{\mu\nu}(z)v^\nu) = \sum_{\pi,\nu} \partial_\pi g_{\mu\nu}(z)v^\pi v^\nu + \sum_\nu g_{\mu\nu}(z)Dv^\nu$. Multiplying the equation by $g^{\rho\mu}$, og

$$\begin{aligned} Dv^\rho &= -\sum_{\pi,\mu,\nu} g^{\rho\mu}\partial_\pi g_{\mu\nu}v^\pi v^\nu + (1/2)\sum_{\pi,\mu,\nu} g^{\rho\mu}\partial_\mu g_{\pi\nu}v^\pi v^\nu \\ &= -\sum_{\pi,\nu}\{{}^{\rho}_{\pi\nu}\}v^\pi v^\nu, \end{aligned} \tag{8.7}$$

where

$$\begin{aligned} \{{}^{\rho}_{\pi\nu}\} &= \sum_\mu g^{\rho\mu}[\pi\nu,\mu], \\ 2[\pi\nu,\mu] &= \partial_\nu g_{\pi\mu} + \partial_\pi g_{\nu\mu} - \partial_\mu g_{\pi\nu}. \end{aligned} \tag{8.8}$$

The $[\pi\nu,\mu]$ are *Christoffel symbols of the first kind*, and the $\{{}^{\rho}_{\pi\nu}\}$ are *Christoffel symbols of the second kind*.

If w and v are vector fields, we define

$$(D_wv)^\rho = \sum_\pi w^\pi\partial_\pi v^\rho + \sum_{\pi,\nu}\{{}^{\rho}_{\pi\nu}\}w^\pi v^\nu, \tag{8.9}$$

provided that the derivative exists. Since $\sum_\pi w^\pi\partial_\pi$ is just the usual directional derivative along a curve with tangent w, the v^μ need be defined and differentiable only on such a curve.

Eq.(8.9) is valid in any chart. One can prove that the $(D_w v)^\rho$ are the components of a vector field, which we denote by $D_w v$ and call the *covariant derivative of v along w*. Eq.(8.7) is then equivalent to $D_v v = 0$: the covariant derivative of v along v vanishes. One calls (8.7), or $D_v v = 0$, the *geodesic equation*. The solutions of the geodesic equation – the curves to which v is tangential – are *geodesics* (of the spacetime metric g). We have therefore shown, for our particular spacetime metric, that the paths of particles subject only to gravitational forces are geodesics.

We shall not make much use of the covariant derivative until Chapter IX. We have mentioned it here because it arises so naturally in the description of particle motion. It is essential in the development of geometrical theories of gravity; a brief account of its properties is given in Appendix D.

Notes

[1] The use of *he* or *she* when addressing the reader is a cause of anguished debate. I shall use both, but not randomly. *He* refers particularly to Joe and *she* to Melissa, whose abilities are very different. You should not be perturbed if the remarks addressed to one of them seem pointless.

[2] We suffer from a surfeit of planetary adjectives. Should one speak of the mercuric or mercurial year, the saturnian, saturnine or saturnalian? Is it permissible to mention the venereal or uranal year? I suggest that we generalize the convention implicit in the naming of the Earth's year. Let us call the planetary years *ymer, yven, year, ymar, yjup, ysat, yura, ynep*, and *yplu*. How fortunate that *y* is both vowel and consonant!

[3] Einstein had a talent for grabbing the wrong ends of important sticks. In the context of their debates on quantum mechanics, Niels Bohr described him as the *Devil* – not in the sense of being personally evil, but as the brilliant advocate of perverse opinions. There is a Gnostic story that the Demiurge, who is the evil creator of the material world, took Jesus up to a high tower, showed him all the nations of the Earth, and said, "All these will be yours if you fall down and worship me." In the Gnostic version, Jesus did. Einstein was doubtless the Demiurge's agent in modern physics.

I am being very disrespectful about one of the great grandfather figures of our age. Einstein was exactly the man for 1905. He took seriously ideas that others only toyed with, and forced physicists to swallow the medicine of relativity and quanta. It was later that he became patron of a stultifying orthodoxy.

[4] The reader learned in the Einstein theory will recall that a geometrical quantity (the Einstein tensor) is assumed to be proportional to the stress-momentum tensor for the not completely convincing reason that the divergence of the former always vanishes, and the divergence of the latter vanishes in special relativity. She may also note that there are perfectly good theories of gravity in which the divergence of the stress-momentum does *not* vanish [Rastall 79a], [Smalley 83]. Einstein's theory should have been a beacon, guiding us to a new kind of physics; instead it became a dogma that for a long time inhibited new ideas. Only now, with the influx of particle physicists and mathematicians, may geometrical physics be coming to fruition.

IV Weight and Energy

1. Mechanics and gravitostatics

The theory that we have developed so far is a small part of gravitostatics: we have considered the motion of single particles and light signals in static or quasistatic gravitational fields. In this chapter, we discuss more general systems in gravitostatics. Later chapters will develop a theory of time-dependent gravitational fields – the postnewtonian theory.

We are going to extend the scope of gravitostatics to include rigid bodies and, to some extent, fluids. As in Newtonian mechanics, there is a choice of strategies: one can regard rigid bodies and fluids as systems of particles, and try to deduce their properties from particle mechanics, or one can treat them as autonomous entities. The first alternative has an attractive, logical sparseness but no profound, physical justification; one cannot really believe that material bodies are made of classical particles. We therefore employ either method, as convenient.

Rigid-body and fluid mechanics are large subjects, which we shall not attempt to treat systematically. Instead, we try to clarify the differences between the Newtonian theory and gravitostatics in some special cases – mainly in relation to the ideas of weight and energy. The topics are chosen partly for their intrinsic interest, but also with an eye to the later development of the postnewtonian theory. The reader will meet once more old friends from her engineering mechanics courses – "light" rods, "inextensible" strings, etc. – and the level of rigour is very similar. We might have called the chapter *An introduction to gravitational engineering*.

A word of warning is necessary. The paths followed in this chapter, particularly in Section 5, are sometimes dead ends. We make a plausible assumption, see where it leads, and then backtrack if necessary. The aim is to give some idea of how one actually does physics. If such inefficient route finding is repugnant, the reader may tiptoe through to the less twisted trails of Chapter V.

2. Static equilibrium

In Newtonian mechanics, a system of particles is in static equilibrium only if the force on each particle vanishes. The science of statics consists in writing this condition in more convenient forms, and applying it to ever more complicated examples. In gravitostatics, we may anticipate that the condition for equilibrium is again that the force on each particle vanishes. However, we shall find that complications arise when one attempts to develop other forms of the equilibrium condition, particularly for rigid bodies.

The momentum of a single particle is constant if the force $\boldsymbol{F}$ that acts on it is zero ($\boldsymbol{F} = \boldsymbol{W} + \boldsymbol{N} = 0$ in III(2.4)). Hence a necessary condition for equilibirum (i.e. for the particle to remain at rest) is that $\boldsymbol{F} = 0$, which implies that

$$\boldsymbol{F} \cdot \boldsymbol{z} = \sum_m F^m z^m = 0 \tag{2.1}$$

for any displacement $\boldsymbol{z} = (z^1,\ z^2\ , z^3)$. Both $\boldsymbol{F}$ and $\boldsymbol{z}$ are regarded as Newtonian quantities although, as shown in the last chapter, $\boldsymbol{F}$ is also the empirical force if we choose the parameters $\tau = 1$, $\lambda = -1$, $\mu = 3$, as in III(3.6). One usually calls $\boldsymbol{z}$ a *virtual displacement* – which means that $\boldsymbol{z}$ is regarded as a small, possible displacement. In the present case, $\boldsymbol{z}$ is *any* small change in the ϕ_0 coordinates of the particle (i.e. the coordinates in a ϕ_0 frame). In Newtonian mechanics, we find the conditions for equilibrium of a mechanical system (which may consist of particles or other material bodies) by generalizing (2.1). We assume that the forces acting on the system may be divided into *external forces* $\boldsymbol{F}_\alpha$, $\alpha \in \{1, \ldots, n\}$, and *internal forces*. Under certain restrictions, a necessary condition for the system to be in equilibrium is that

$$\sum_\alpha \boldsymbol{F}_\alpha \cdot \boldsymbol{z}_\alpha = 0 \tag{2.2}$$

for any virtual displacements $\boldsymbol{z}_\alpha$ of the points of application of the forces $\boldsymbol{F}_\alpha$. The $\boldsymbol{z}_\alpha$ are again arbitrary, small, possible displacements – where *possible* means that, for example, they do not distort rigid bodies or violate imposed boundary conditions. One of the restrictions is that the $\boldsymbol{F}_\alpha$ be independent of the $\boldsymbol{z}_\alpha$ (which means that frictional forces are excluded). Another restriction is that, in any virtual displacement of the system, the internal forces do no work. (We recall "Newton's Third Law", which ensures that very special, two-body forces satisfy this condition.)

In gravitostatics also, we may consider (2.2) to be the condition for static equilibrium, provided that we now interpret $\boldsymbol{F}_\alpha$ and $\boldsymbol{z}_\alpha$ as Newtonian quantities. We must still, of course, impose the restrictions that the $\boldsymbol{F}_\alpha$ be independent of the $\boldsymbol{z}_\alpha$, and that the internal forces do no work. We note that if the $\boldsymbol{z}_\alpha$ can be chosen to be all equal to an arbitrary, small $\boldsymbol{z}_1$, then (2.2) implies that a necessary condition for equilibrium is that the total force on the system vanish: $\sum_\alpha \boldsymbol{F}_\alpha = 0$.

Weight of a rod

As a first application of (2.2), we consider the equilibrium of a thin, uniform, rigid rod in a gravitational field. By *thin* we mean that the transverse dimensions of the rod are very small in comparison with its length; by *uniform* that its properties – the mass per unit length, cross section, etc. – are the same everywhere along its length; by *rigid* that the rod does not bend (can you make that precise?), and its empirical length l_E is constant: l_E does not depend significantly on the gravitational field or the applied forces.

We introduce orthonormal vectors $\boldsymbol{i}_m$ parallel to the coordinate axes of the ϕ_0 frame x. The gravitational potential is static and independent of the coordinates x^1 and x^2, and $\partial_3\phi > 0$. We refer to the directions of $\boldsymbol{i}_3$ and $-\boldsymbol{i}_3$ as *up* and *down*, respectively. In performing the calculation, we do not assume that the parameters λ, τ, μ are given by III(3.6). This is useful later in the chapter when we discuss other ways of deriving their values.

The rod is at rest in x with its length parallel to $\boldsymbol{i}_1$. A force $F\boldsymbol{i}_1$ is applied to the end of the rod with the greater value of the coordinate x^1, and a force $H\boldsymbol{i}_1$ to the other end; there is a downward gravitational force $\boldsymbol{J} = -J\boldsymbol{i}_3$ (cf. III(4.6)), and a sustaining force $\boldsymbol{S} = S\boldsymbol{i}_3$ that holds the rod stationary.

If the rod is displaced by a small amount $z\boldsymbol{i}_1$, then all the $\boldsymbol{z}_\alpha$ are equal to $z\boldsymbol{i}_1$ and (2.2) becomes $F + H = 0$. If the rod is displaced by a small amount $z\boldsymbol{i}_3$, its length l changes by $\delta l \approx -\lambda l \delta\phi \approx -\lambda l z \partial_3\phi$ (use II(7.4), and note that the empirical length $l_E = l_0$ is constant). If the rod remains parallel to $\boldsymbol{i}_1$ during this displacement, and its centre is not displaced in the $\boldsymbol{i}_1$ and $\boldsymbol{i}_2$ directions, then the displacements of its ends are $z\boldsymbol{i}_3 \pm (1/2)\lambda l z \partial_3\phi \boldsymbol{i}_1$, where the lower sign refers to the end with the greater value of x^1, at which the force $F\boldsymbol{i}_1$ acts. From (2.2) again, it follows that $-F\lambda l z\partial_3\phi + H\lambda l z \partial_3\phi + 2(S - J)z = 0$, and hence $S = J + F\lambda l \partial_3\phi$.

The result that we have just derived is very peculiar. It implies that if $F\lambda \neq 0$, then $\boldsymbol{S} + \boldsymbol{J} \neq 0$: the total force does not vanish even though the system is in equilibrium. We must suspect that something is wrong – and it is. Our error was to use (2.2), which is valid only if the internal forces do no work. It may appear obvious that they do not, because the rod has a constant empirical length. However, the argument was in terms of *Newtonian* quantities; the Newtonian length depends on the potential, and as it changes the internal forces do work. To rectify matters, we make the plausible assumption that the internal force acting on an end portion of the rod should balance the horizontal external force of magnitude F (we may regard an end portion as a kind of particle, which is in equilibrium when the total force on it vanishes). The work done by the internal forces then balances that done by the horizontal forces, the term $\lambda F l \partial_3 \phi$ disappears, and $S = J$ – or in vector form

$$\boldsymbol{S} = -\boldsymbol{J}. \tag{2.3}$$

The moral of this story is that one must be cautious when dealing with rigid bodies. A body that is rigid in terms of empirical measurements does not in general have a constant Newtonian size.

Eq.(2.3) shows that, as in the Newtonian theory, the gravitational force on the rod is the negative of the sustaining force; either can be called the *weight* of the system. For many laboratory objects, it may be convenient to regard the weight as the negative of the sustaining force, if this can be directly measured. For most astronomical objects, it is better to regard the weight as the gravitational force – one cannot suspend the Earth in a uniform gravitational field, for example.

Although we have shown that the sustaining force is the negative of the gravitational force on the rod, we still do not know whether either of them depends on the stress. To resolve the question, we need a more physical argument. We must somehow compare the rod in its stressed and unstressed states.

We consider the same configuration as before, but now write the sustaining force as $\boldsymbol{S}_0$ when the rod is unstressed ($F = 0$) and $\boldsymbol{S}$ when the forces $F\boldsymbol{i}_1$ and $-F\boldsymbol{i}_1$ are applied to the ends. We begin with the rod in its unstressed state, and consider two sequences of operations.

(a) First slowly displace the rod by $z\boldsymbol{i}_3$, then increase the force on the ends from zero to F.

(b) First increase the force on the ends from zero to F, then slowly displace the rod by $z\boldsymbol{i}_3$.

(By *slowly* we mean *quasistatically* – the idea is to ensure that the kinetic energy is negligible.)

The work done by the applied forces in (a) and (b) must be the same – otherwise one can violate the conservation of energy by performing first (a) and then the inverse of (b). Since the applied forces are the sustaining force and the forces at the ends, it follows that $S_0 z = Sz - \lambda F l z \partial_3 \phi$. Applying (2.3) to the stressed and unstressed rod, we find that $\boldsymbol{J}_0 = -\boldsymbol{S}_0$, $\boldsymbol{J} = -\boldsymbol{S}$, and hence the gravitational forces (or weights) are related by

$$\boldsymbol{J} = \boldsymbol{J}_0 - \lambda F l \nabla \phi. \tag{2.4}$$

This tells us how the weights of the stressed and unstressed rods are related. Note that when $\lambda = -1$ and the rod is stretched ($F > 0$), the magnitude of the weight is decreased.

Exercise 2.1 Apply (2.4) to a steel bar of length 1m. Choose F to be close to the force that breaks the bar, and estimate the fractional change in J. ■

Weight of a box

In the next section, we calculate the weight of a box full of gas, and we shall need to know the weight of the box alone. It is easily found by generalizing the argument just used for a rod. We consider a 3 dimensional, rectangular box with thin, uniform, rigid walls (we leave it to the reader to write down reasonable definitions of these terms). From II(7.4), the Newtonian volume B of the box is related to the empirical volume B_E by $B = B_E e^{-3\lambda\psi}$, $\psi = \phi - \phi_0$. The assumption that the walls are thin implies that we need not distinguish between inside and outside volume. The assumption of rigidity implies that B_E is constant – it does not depend significantly on the gravitational field or the applied forces. We are supposing that the volume is a local quantity, which means that ϕ must not change appreciably over the volume, and hence that $|\nabla\phi|$ is small. We take $\nabla\phi$ to be almost constant throughout the box, and $\nabla\phi = i_3\partial_3\phi$ with $\partial_3\phi > 0$, as before.

We assume that a pressure force acts on the walls of the box in the direction of the outward normal. The pressure (the magnitude of the force per unit area) is p in Newtonian units; like ϕ, it does not vary appreciably from point to point in the box.

We find the weight of the box (the negative of the sustaining force $\boldsymbol{S}$) by an argument exactly similar to that used for a rod in the derivation of (2.4). We suppose that the box is suspended in the gravitational field, and we write the sustaining force as $\boldsymbol{S}_0$ when it is unstressed (zero pressure), and $\boldsymbol{S}$ when the pressure is p. We begin with the unstressed box, and consider two sequences of operations.

(a) First slowly displace the box by zi_3, then increase the pressure from zero to p.

(b) First increase the pressure from zero to p, then slowly displace the box by zi_3.

By the same argument as before, the work done by the applied forces in (a) and (b) must be the same. The applied forces are the sustaining force and the pressure forces. The change in the volume B of the box is $-3\lambda Bz\partial_3\phi$, and the work done by the pressure forces in (b) is p times the change in volume. We have therefore shown that $S_0 z = Sz - 3\lambda pBz\partial_3\phi$, or

$$\boldsymbol{S} = \boldsymbol{S}_0 + 3\lambda pB\nabla\phi. \tag{2.5}$$

For the value $\lambda = -1$, one has $S < S_0$, and the magnitude of the weight of box is less when it is stressed by pressure forces from inside (which corresponds to $p > 0$). One can also take $p < 0$, corresponding to pressure forces compressing the box from outside, and one then has a bigger weight: $S > S_0$ when $\lambda = -1$. As in (2.3), we have identified the weight (the gravitational force acting on the box) with the negative of the sustaining force.

The results that we have derived may seem strange. Rigid bodies are defined as having constant empirical dimensions but, as they move in the Euclidean space of the theory, their Newtonian (or Euclidean) dimensions change, and forces do work on them because of these changes. As a consequence, the weight of a rigid body is stress-dependent. We may regard this as the price paid for using Euclidean geometry (in terms of which the body's size is not constant).

3. Weight and energy

The force on a body in Newtonian mechanics is equal to the inertial mass multiplied by the acceleration of the centre of mass. The experiments of Eötvös showed that the weight of a solid body is proportional to its inertial mass, the constant of proportionality being the same for a variety of materials [Misner 73]. Two bodies of different materials but equal weight are suspended from a horizontal beam. As the Earth rotates, the bodies move in a circle whose centre is on the Earth's axis. The suspension exerts forces on the bodies that keep them on their circular path, and the measurements show that these forces are equal. Since the bodies have the same acceleration, it follows that their inertial masses are equal. Later, more accurate experiments by Dicke and Braginsky have compared the accelerations of bodies of different materials in the Sun's gravitational field. Their results are compatible with the accelerations being equal. Small discrepancies in Eötvös' experiments were one of the motivations for recent attempts – so far unsuccessful – to find a short-range component in the gravitational force.

The magnitude of the gravitational force on a *stationary* body, divided by the magnitude $|\nabla\phi|$ of the gravitational field in its neighbourhood, is called the *passive gravitational energy* of the body. The *passive gravitational mass* of the body is c^{-2} times the passive gravitational energy.[1] The definitions must be modified in an obvious way if $|\nabla\phi|$ varies significantly over the volume of the body. Experiments of the Eötvös type establish the equality of inertial and passive gravitational mass. However, the gravitational energy of laboratory-size bodies is very much less than their rest energy, and the experiments are insufficiently precise to show whether gravitational energy contributes in the same way (or at all) to the passive gravitational and inertial masses.

In astronomical bodies, a significant fraction of the energy may be gravitational, and one can try to measure its effect on the bodies' motions. One may ask, for example, whether the Earth and the Moon, which have different fractions of gravitational energy, fall towards the Sun at different rates. If they do, the Earth-Moon orbit will be distorted (the *Nordtvedt effect*). Very precise measurements of the Moon's orbit have been made, using a laser on Earth and a reflector on the Moon. They are consistent with there being no distortion: the Earth and Moon fall at the same rate.

Exercise 3.1 Make *rough* estimates of the gravitational energy of the Earth and the Moon, and of the difference between their accelerations towards the Sun if gravitational energy contributes to their inertial masses but *not* to their passive gravitational masses. Also estimate the size of the distortion of the Earth-Moon orbit. ■

Weight of a box of gas

In the last section, we calculated the weight of an empty, stressed box. Now we consider a box that contains ideal gas – a large number N of weakly interacting particles, each of mass m. The stress in the box is due to the pressure of the gas.

The relationship between the weight and the total energy of the box of gas is not obvious. The gravitational force on a particle of given energy depends on its speed, by III(4.6); the total gravitational force, found by summing over all the particles, therefore depends on the velocity distribution. We must also take account of the weight of the box. The weight of gas plus box is the negative of the sustaining force that holds the box of gas at rest in the gravitational field. We find the weight of the gas alone by subtracting the weight of the box.

We label the particles in the box with an index $b \in \{1, \ldots, N\}$, and take V_b and E_b to be the speed and energy, respectively, of particle b. The box has volume B. In the lowest, Newtonian approximation – which is sufficient here – the pressure of the gas in the *absence* of a gravitational field is[2]

$$p = (3B)^{-1} \sum_b m V_b^2, \tag{3.1}$$

where $V_b^2 = |\boldsymbol{V}_b|^2$, and is not to be confused with the 2 component of the velocity $\boldsymbol{V}_b$.

In the presence of the gravitational field, the pressure is greater at the bottom of the box than at the top: it is this pressure difference that holds up the gas. The gravitational force on an individual particle is

$$\boldsymbol{W} = -E\big[1 + (\tau - \lambda - 1)V^2 c_N^{-2}\big] \nabla\psi(\boldsymbol{Z}, \,.), \tag{3.2}$$

where we have used III(2.3), III(2.5), and the Newtonian limit condition III(3.1), and written $V^2 = |\boldsymbol{V}|^2$. We assume that $\nabla\psi(\boldsymbol{Z}, \,.) = \alpha i_3$ is constant throughout the box, and that V is small so that $Ec_N^{-2} \approx m$. From III(2.4), the equation of motion of a particle is $D\boldsymbol{P} = \boldsymbol{W} + \boldsymbol{N}$, where $\boldsymbol{N}$ is the nongravitational force acting on it. Labeling the particles with a suffix b again, we write $D\boldsymbol{P}_b = \boldsymbol{W}_b + \boldsymbol{N}_b$. Summing over all the particles, og

$$\begin{aligned}\sum_b \boldsymbol{N}_b = -\sum_b \boldsymbol{W}_b = -\boldsymbol{J} &= \sum_b \big[E_b + m(\tau - \lambda - 1)V_b^2\big]\alpha i_3 \\ &= \big[\mathcal{E} + 3(\tau - \lambda - 1)Bp\big]\alpha i_3,\end{aligned} \tag{3.3}$$

where $\mathcal{E} = \sum_b E_b$, p is given by (3.1), and $\boldsymbol{J}$ is the total gravitational force on the particles.

The force $-\boldsymbol{J}$ is that required to sustain the particles. In addition, there is the force required to sustain the box, which we calculate from (2.5) with $\nabla\phi = \alpha i_3$. The magnitude of the total weight of the system (box plus particles) is therefore $S_0 + 3\lambda pB\alpha + \big[\mathcal{E} + 3(\tau - \lambda - 1)Bp\big]\alpha = S_0 + \big[\mathcal{E} + 3(\tau - 1)Bp\big]\alpha$. If $\tau = 1$, this becomes $S_0 + \mathcal{E}\alpha$ – which is the weight of the unstressed box plus the weight of *stationary* particles whose total energy is the same as the total energy $\mathcal{E}$ of the particles in the box. In other words, when $\tau = 1$ the kinetic energy contributes to the weight in the same way as the energy of a stationary particle. We emphasize that we are concerned with a system whose centre of mass is at rest in a static gravitational field. Putting $\tau = -\lambda = 1$ in (3.3), we find that the passive gravitational energy in Newtonian units *of the particles only* is

$$J/\alpha = \mathcal{E} + \sum_b mV_b^2 = (\mathcal{E} + 3Bp). \tag{3.4}$$

4. Energy of the gravitational field

The gravitational field is, so far, no more than a convenient device for calculating gravitational forces. We recall that the electric field played a similar role in electrostatics, but that it soon found other uses – first as a place to store the electric potential energy, and then in the description of electromagnetic waves.[3] A similar development in Newtonian gravity was halted when it was shown that the energy density of the gravitational field is negative. In a strictly static theory this would not matter, but as soon as there is any time dependence a system will presumably evolve to states of lower energy. Since there is no state of *lowest* energy – the gravitational energy can take arbitrary negative values – the system cannot attain equilibrium and singularities will develop. We must now see whether the same difficulty arises in gravitostatics.

Let us first recall some earlier terminology and results. We have defined the gravitational field to be $-\nabla\phi$, where ϕ is the dimensionless potential. The gravitational force on a stationary particle of energy E is $-E\nabla\phi(\boldsymbol{Z},\ .)$, where $\boldsymbol{Z} = (Z^1,\ Z^2,\ Z^3)$ are the coordinates of the particle in a ϕ_0 frame. For a particle of velocity $\boldsymbol{V} = D\boldsymbol{Z}$, the gravitational force is $-E\big[1+V^2c_N^{-2}\big]\nabla\phi(\boldsymbol{Z},\ .)$, from III(4.6) with $V^2 = |\boldsymbol{V}|^2$ (we are using the standard values III(3.6) of the parameters λ, τ, and μ). When ϕ is static ($\partial_t\phi = 0$) and only gravitational forces act on the particle, the quantity $E = mc_N^2\gamma$ is a constant of the motion. One calls E the *energy* of the particle. It is reprehensible, but common, to use the same name even when E is not constant.

We begin with a trivial problem in Newtonian gravity. A particle of mass m is fixed; a second, equal particle is brought slowly from a great distance, and comes to rest at a distance l from the first. (We omit E subscripts because in Newtonian theory all quantities are empirical.) In order for the second particle to move slowly, a force $\boldsymbol{N}$ must act on it that balances the gravitational pull due to the first particle. The magnitude of this force at radial distance r is Gm^2r^{-2}, and the total work that it does on the second particle is $\int_\infty^l Gm^2r^{-2}dr = -Gm^2l^{-1}$. One defines $-Gm^2l^{-1}$ to be the increase in *potential energy* of the system, so that the work done on the system is the increase in its potential energy.

The preceding calculation is valid as a first approximation in gravitostatics, provided that m is not large and l not small. The potential due to one particle at the position of the other is then small, and one can replace m by m_E in the integral. The total energy of the system when the particles are widely separated is $2m_Ec^2$ (we choose the ϕ_0 frame so that Newtonian and empirical energy are the same in this case), and in the final state the Newtonian energy is

$$\mathcal{E} = 2m_Ec^2 - Gm_E^2l^{-1} = 2m_Ec^2 + m_Ec^2\psi, \tag{4.1}$$

where $\psi = \phi(\boldsymbol{Z}) - \phi_0$, and $\phi(\boldsymbol{Z})$ is the potential at the final position of either particle. In terms of the energy $E = mc_N^2 = m_Ec^2e^\psi \approx m_Ec^2(1+\psi)$ of each particle (cf. III(4.3)), og

$$\mathcal{E} = 2E + Gm_E^2l^{-1} = 2E - m_Ec^2\psi. \tag{4.2}$$

Note that the potential energy term appears with different signs in (4.1) and (4.2). In (4.2), part of the work done by the force has gone to change the energy of the individual particles, and the remaining term is positive. We shall find that the latter can be interpreted in terms of a positive, gravitational energy density.

We now refine and generalize the preceding argument so that it applies to an arbitrary, bounded, static mass distribution. The method is similar to that used by Helmholtz for the electrostatic field ([Whittaker 51], pp. 217–218). We first derive an expression, eq.(4.3), for the rate at which a force does work on a particle, and then calculate the work done in assembling the mass distribution from an initial state in which the matter is widely dispersed (the traditional phrase is that the matter is *brought up from infinity* – a Godvomit).

From III(4.5), the momentum of a particle of mass m and velocity $\boldsymbol{V} = D\boldsymbol{Z}$ is $\boldsymbol{P} = m\gamma\boldsymbol{V}$. If a gravitational force $\boldsymbol{W}$ and a non-gravitational force $\boldsymbol{N}$ act on the particle, then $D\boldsymbol{P} = \boldsymbol{W} + \boldsymbol{N}$, from III(2.4). The rate at which the force $\boldsymbol{N}$ does work on the particle is $\boldsymbol{N}\cdot\boldsymbol{V} = -\boldsymbol{W}\cdot\boldsymbol{V} + D\boldsymbol{P}\cdot\boldsymbol{V}$. To calculate the last term, one writes III(2.6) as $\boldsymbol{P}\cdot\boldsymbol{P} = E^2c_N^{-2} - m^2c_N^2$, differentiates with respect to t, uses III(4.1) and III(4.2), and gets $D\boldsymbol{P}\cdot\boldsymbol{V} = DE - E(1 + V^2c_N^{-2})D\phi(\boldsymbol{Z}, .)$. One has

$D\phi(\boldsymbol{Z},\,.) = \boldsymbol{V}\cdot\nabla\phi(\boldsymbol{Z},\,.) + \partial_t\phi(\boldsymbol{Z},\,.)$, and hence

$$\boldsymbol{N}\cdot\boldsymbol{V} = DE - E(1+V^2c_N^{-2})\partial_t\phi(\boldsymbol{Z},\,.). \tag{4.3}$$

We are now ready to calculate the work done by the nongravitational forces in assembling a matter distribution with energy density ϵ. The final matter distribution is assumed to be static, so that $\partial_t\epsilon = 0$, and the potential due to ϵ satisfies the Poisson equation III(2.7) with the boundary condition that $\phi(\boldsymbol{x}) \to \phi_0$ as $|\boldsymbol{x}| \to \infty$. If we define $\phi_{(s)} = s\phi + (1-s)\phi_0$ for any real number s, then $\phi_{(s)}$ satisfies III(2.7) with ϵ replaced by $s\epsilon$, and $\phi_{(s)}(\boldsymbol{x}) \to \phi_0$ as $|\boldsymbol{x}| \to \infty$. Thus $\phi_{(s)}$ is the potential corresponding to the energy density $s\epsilon$.

We suppose that the matter distribution is assembled very slowly by bringing small particles, one at a time, from far away where the potential is very close to ϕ_0. We assume that at a certain stage in this process the energy density of the matter distribution is $s\epsilon$, where $0 \le s \le 1$. To find the work done in increasing the energy density from $s\epsilon$ to $(s+\delta s)\epsilon$, we apply (4.3) to each particle that we bring from far away. If E is the energy of the particle, then $\partial_t\phi$ is proportional to E (since the change in ϕ is due to the motion of the particle). The last term in (4.3) is therefore proportional to E^2, and can be made negligible by choosing E to be small. It follows that, to first order in δs, the work done in increasing the energy density from $s\epsilon$ to $(s+\delta s)\epsilon$ is equal to the increase in energy of the matter that is moved. Note that, throughout this calculation, we neglect terms quadratic in δs.

In changing the energy density from $s\epsilon$ to $(s+\delta s)\epsilon$, we change the potential from $\phi_{(s)}$ to $\phi_{(s+\delta s)} = \phi_{(s)} + (\phi - \phi_0)\delta s$, where we again neglect terms of order $(\delta s)^2$. During this change, the matter already assembled remains fixed in the ϕ_0 frame so that no work is done on it. Hence the *empirical* energy per unit ϕ_0 3-volume of the already assembled matter, which we denote by ζ, is constant during the change.[4] The *Newtonian* energy density of the already assembled matter is initially $s\epsilon = \zeta\exp(\phi_{(s)} - \phi_0)$, from III(4.3), and becomes $\zeta\exp(\phi_{(s+\delta s)} - \phi_0) = s\epsilon\exp(\phi_{(s+\delta s)} - \phi_{(s)}) = s\epsilon[1 + (\phi - \phi_0)\delta s]$. The increase in Newtonian energy of the already assembled matter when the potential is changed from $\phi_{(s)}$ to $\phi_{(s+\delta s)}$ is therefore $s\delta s \int \epsilon(\boldsymbol{x})(\phi(\boldsymbol{x}) - \phi_0)\, d^3x$, where the integral is over the whole matter distribution. (This quantity *plus* the energy of the matter brought up is of course $\delta s \int \epsilon(\boldsymbol{x})\, d^3x$).

We are assuming conservation of energy – that the increase in energy of the system is equal to the work done by the forces that act on it. We also assume that the energy of the system is the energy of the matter distribution plus the energy of the gravitational field. We have shown that when the energy density changes from $s\epsilon$ to $(s+\delta s)\epsilon$, the work done is equal to the increase in energy of the matter that is moved. The energy of the already assembled matter increases by

$s\delta s \int \epsilon(\boldsymbol{x})(\phi(\boldsymbol{x}) - \phi_0)\, d^3x$, and the energy of the gravitational field must therefore decrease by the same amount. Integrating from $s = 0$ to $s = 1$, we find $\mathcal{E}_G$, the increase in the energy of the gravitational field on bringing the matter from a dispersed state to the state where the energy density is ϵ:

$$\mathcal{E}_G = -\tfrac{1}{2} \int \epsilon(\boldsymbol{x})(\phi(\boldsymbol{x}) - \phi_0)\, d^3x. \tag{4.4}$$

We can express $\mathcal{E}_G$ in terms of ϕ only by using the field equation $\nabla^2\phi = 4\pi k\epsilon$, eq.III(2.7). Since the matter distribution is bounded, the term $\phi(\boldsymbol{x}) - \phi_0$ is of order $|\boldsymbol{x}|^{-1}$ as $|\boldsymbol{x}| \to \infty$. The identity $\psi\nabla^2\psi = \text{div}\,(\psi\nabla\psi) - |\nabla\psi|^2$ and the divergence theorem then give $\mathcal{E}_G = \int \epsilon_G(\boldsymbol{x})\, d^3x$, where

$$\epsilon_G = (8\pi k)^{-1}|\nabla\phi(\boldsymbol{x})|^2 \tag{4.5}$$

is the *energy density* of the gravitational field (the ϕ_0 energy per unit ϕ_0 3-volume).

The expression (4.5) for ϵ_G is positive definite, and is exactly similar to the corresponding expression for the electric field. The difference from the Newtonian calculation, as we previously showed in the two-particle example, is that the energy of the already assembled matter depends on the potential, and changes as more matter is added.

We express the total energy $\mathcal{E}$ of the system as the sum of the energy of the matter $\mathcal{E}_M = \int \epsilon(\boldsymbol{x})\, d^3x$ and the energy of the gravitational field $\mathcal{E}_G$:

$$\mathcal{E} = \mathcal{E}_M + \mathcal{E}_G = \int \Big[\epsilon(\boldsymbol{x}) + \epsilon_G(\boldsymbol{x})\Big]\, d^3x. \tag{4.6}$$

Writing $\psi = \phi - \phi_0$, noting that $\epsilon e^{-\psi/2} = \epsilon - \epsilon\psi/2 + O(\psi^2)$ as $\psi \to 0$, and using (4.4), og

$$\mathcal{E} = \int \epsilon e^{-\psi/2}(\boldsymbol{x})\, d^3x \tag{4.7}$$

to the accuracy of the preceding calculations (where terms of order ψ^2 are neglected).

One may ask why the energy density of the gravitational field is taken to be ϵ_G rather than the integrand $\epsilon_G^* = -(1/2)\epsilon(\boldsymbol{x})(\phi(\boldsymbol{x}) - \phi_0)$ of (4.4), which differs from ϵ_G by a divergence. A good feature of ϵ_G is that it vanishes where the gravitational field vanishes. As for ϵ_G^*, it vanishes where ϵ vanishes, even though the gravitational field may be nonzero there. It is perhaps plausible to regard the integral of ϵ_G^* over the volume of a material body as being the contribution of the gravitational field to the energy of the body. There is no compelling reason for preferring either of the expressions for the energy density, and in the next section we consider both possibilities.

Exercise 4.1 Calculate ϵ_G at the surfaces of the Earth, the Sun, a white dwarf, and a neutron star. Compare your results with the energy density of water at standard temperature and pressure. [Assume that the masses of the white dwarf and the neutron star are of the same order as the mass of the Sun (about 2×10^{30} kg). The radius of a white dwarf is about 10^4 km, and that of a neutron star is about 10 km.] ■

5. Gravitational energy and active gravitational mass

We showed in the last section that the total energy of a static system is $\mathcal{E} = \mathcal{E}_M + \mathcal{E}_G$, where $\mathcal{E}_M = \int \epsilon(\boldsymbol{x})\, d^3x$ is the energy of the sources and $\mathcal{E}_G = \int \epsilon_G(\boldsymbol{x})\, d^3x$ is that of the gravitational field. Equivalently, we have $\mathcal{E} = \int \epsilon e^{-\psi/2}(\boldsymbol{x})\, d^3x$ as in (4.7). However, when developing gravitostatics we assumed that the active gravitational mass of a system is proportional to the total Newtonian energy of the sources *without* the gravitational energy; the source term in the Poisson equation III(2.7) is $4\pi k\epsilon$, *not* $4\pi k(\epsilon+\epsilon_G)$ or $4\pi k\epsilon e^{-\psi/2}$, and the gravitational energy does not contribute to the active gravitational mass.

Although it might be aesthetically pleasing if gravitational and nongravitational energy contributed in the same way to the active gravitational mass, there is no compelling physical reason to believe it. In the end, all that one can do is calculate the consequences of the assumption and compare the results with observation.

We first consider the possibility that, instead of III(2.7), the field equation is

$$\nabla^2\psi = 4\pi k_N \epsilon e^{-\psi/2} \approx 4\pi k_N(\epsilon + \epsilon_G^*), \tag{5.1}$$

where $\psi = \phi - \phi_0$, as before. If the sources are spatially bounded, the solution of (5.1) at large spatial distances is of the form III(3.2). Imposing the condition that $\psi(\boldsymbol{x}) \to 0$ as $r = |\boldsymbol{x}| \to \infty$, og $\psi(\boldsymbol{x}) = -k\mathcal{E}'/r + O(r^{-2})$ as $r \to \infty$, where $k = k_E$ and

$$\mathcal{E}' = \int \epsilon(\boldsymbol{x}) \exp\left[(\lambda - 2\tau + \mu - \tfrac{1}{2})\psi(\boldsymbol{x})\right] d^3x. \tag{5.2}$$

Eq.(5.2) differs from III(3.3) by the term $-1/2$ in the parenthesis.

The assumption that the active gravitational mass is proportional to the energy *including the gravitational energy* means that $\mathcal{E}' = \mathcal{E}$, where $\mathcal{E}$ is given by (4.7), and hence that $\lambda - 2\tau + \mu = 0$ as in III(3.4).

If the Lagrangian of a particle is III(2.1), then the values of the parameters τ, λ, and μ are unchanged, and the predicted results for the red shift, bending of light, and perihelion shift of a small particle are the same as in Chapter III. However, if the particle is not small, we may expect that (5.1) will predict different results from III(2.7). If we consider a binary star, for example, the integral (5.2) taken over the volume of one star depends on the position of the other. Consequently, the potential $\psi(\boldsymbol{x})$ due to the first star contains terms proportional to r^{-2}. These terms will be calculated in Chapter VIII.

Instead of taking the Lagrangian to be $-mc_N^2\gamma^{-1}$ (we are dropping the arguments of functions with wild abandon!), we might choose $-mc_N^2 e^{-\psi/2}\gamma^{-1}$. That is, we might replace mc_N^2 by $mc_N^2 e^{-\psi/2}$, just as we replaced ϵ by $\epsilon e^{-\psi/2}$ to get (4.7). The next two exercises are concerned with such possibilities.

Exercise 5.1 Assume that the Lagrangian of a particle is $-mc_N^2 e^{\alpha\psi}\gamma^{-1}$, where α is a constant. Use the arguments of Chapter III to find the parameters τ, λ, and μ, and hence calculate the red shift, bending of light, and relativistic perihelion advance. ■

Exercise 5.2 In deriving (4.4), we assumed the standard values $\tau = 1$, $\lambda = -1$, $\mu = 3$, of III(3.6). Repeat the derivation for any τ, λ, μ, and for the more general Lagrangian $-mc_N^2 e^{\alpha\psi}\gamma^{-1}$. ■

Another way of modifying the field equation III(2.7) is to replace ϵ by $\epsilon + \epsilon_G$ on the right-hand side:

$$\nabla^2\psi = 4\pi k(\epsilon + \epsilon_G), \tag{5.3}$$

where ϵ_G is given by (4.5). We choose the parameters λ, τ, μ as in III(3.6) (which implies that $k = k_N$). There seems to be an inconsistency, because we used III(2.7) in deriving (4.5). However, if $\epsilon_G \ll \epsilon$, which is true almost everywhere in our galaxy, the resulting error in the expression for ϵ_G is negligible.

The difference between ϵ_G in (5.3) and ϵ_G^* in (5.1) is only a spatial divergence. This does not change the total gravitational energy, but it does change the distribution of gravitational energy – with observable consequences, as we shall see.

To solve (5.3), we write $j = e^{-\psi/2}$. Using (4.5), we find that $\nabla^2 j = -2\pi k j\epsilon$, which reduces to Laplace's equation in a region where ϵ vanishes. In the special case when ϵ is spherically symmetric and bounded (so that $\epsilon(\boldsymbol{x}) = 0$ when $r = |\boldsymbol{x}| > r_0$ for some constant r_0), og $j = j_0 + k\mathcal{E}/2r$ for $r > r_0$, where $\mathcal{E}$ is again given by (4.7) (i.e. $\mathcal{E}$ is the energy of the sources, including the gravitational energy). If $\psi \to 0$ as $r \to \infty$, as we have assumed, og $j_0 = 1$.

If ψ is small, we expand j as a power series, neglect terms in ψ^3, and find that $-\psi/2 + \psi^2/8 = k\mathcal{E}/2\mathrm{r}$. To first approximation, $\psi = -k\mathcal{E}/r$, and substituting this value into the small ψ^2 term og as a second approximation

$$\psi(\boldsymbol{x}) = -k\mathcal{E}/r + (k\mathcal{E}/2r)^2, \tag{5.4}$$

valid when $r > r_0$, and provided that terms in $(\mathcal{E}/r)^3$ are negligible.

The calculation of the redshift in Chapter III is independent of the choice of field equation, and is therefore unchanged. We again assume that the Lagrangian of a particle is III(2.1), and we easily check that the additional term $(k\mathcal{E}/2r)^2$ in (5.4), as compared with III(5.1), makes no appreciable difference to the deflection of light by the Sun or the radar time delay. The perihelion advance is however different from that calculated in Section III.6. The force derived from (5.4) is still radial, so that the angular momentum remains a constant of the motion, and III(6.2) still holds, but the constant A is changed and the relativistic perihelion advance πA becomes

$$\pi A \approx (11\pi/2)(GMmj^{-1}c^{-1})^2, \tag{5.5}$$

which is about 8 percent less than that given by III(6.6).

The relativistic perihelion advance of Mercury as deduced from observation agrees with III(6.6) to less than 1 percent. We must therefore reject the assumption that the active gravitational mass depends on the gravitational energy of the sources in the manner of (5.3). In the next chapter, we adopt another hypothesis – equally acceptable as a generalization of Newtonian theory – that the active and passive gravitational masses are the same.

Although we glibly said that the relativistic perihelion advance is deduced from observation, we should emphasize how difficult this is. Because it rotates, the Sun is not spherically symmetric. Its gravitational quadrupole moment produces a perihelion shift which must be computed, like all the other Newtonian effects, before the residual, relativistic perihelion advance can be found. It had been assumed that the centre of the Sun rotates at much the same rate as the observable, surface region, which would imply that the quadrupole moment and the resulting perihelion shift are small. However, the shape of the Sun was measured by Dicke and Goldenberg in 1966. Their results were interpreted as implying a much larger quadrupole moment, and a relativistic perihelion advance in very good agreement with (5.5)! The results were not confirmed by later observations by Hill, Dicke, and their collaborators.

By measuring the oscillations of the sun ("helioseismology"), one can deduce the rotation speed in the deep interior. Recent measurements show that the Sun rotates fairly uniformly, and that its quadrupole moment gives a negligible contribution to the perihelion advance (see the review articles by I.I. Shapiro and others in [Ashby 89]).

Exercise 5.3 Calculate the motion of two, spherically symmetric bodies with the same inertial and passive gravitational masses, but different active gravitational masses (write the ratio of the active gravitational masses as $1 + \beta$). Assume that Newtonian theory is valid, apart from the difference between the active gravitational masses, and that no forces act on the bodies apart from their mutual gravitational attraction. Describe the orbit both in terms of the relative motion of the particles, and in terms of a Galilean frame [Note that the centre of mass accelerates in the Galilean frame! Does the perihelion advance?] ■

Notes

[1] The definition of passive gravitational mass is analogous to the definition in Section III.2 of active gravitational mass as $\mathcal{E}c^{-2}$, where $\mathcal{E}$ is the active gravitational energy. Since $\nabla\phi$ is in Newtonian units, so is the passive gravitational energy ($\partial_m \phi$ is the derivative of the dimensionless quantity ϕ with respect to the x^m coordinate). However, the passive gravitational mass is *not* in Newtonian units. To get a mass in Newtonian units, we could have divided the force by $c_N^2|\nabla\phi|$, rather than $c^2|\nabla\phi|$.

[2] *A* simple proof is as follows. Consider collisions of the gas particles with a wall whose outward normal is in the x^1 direction (i.e. the gas is on the side of the wall that corresponds to smaller values of x^1). Let there be n_β particles per unit volume whose x^1 velocity component is V^1_β (or more precisely, whose x^1 velocity component is in some small interval that contains V^1_β). If $V^1_\beta < 0$, these particles do not hit the wall; if $V^1_\beta > 0$, the number of them that hit the wall, per unit area and unit time, is $n_\beta V^1_\beta$. The momentum they transfer to the wall is $2mn_\beta(V^1_\beta)^2$ – and this is their contribution to the pressure on the wall. The total pressure p on the wall is $p = \sum_{+\beta} 2mn_\beta(V^1_\beta)^2$, where the sum is over all β for which $V^1_\beta > 0$. For a gas in equilibrium, one can write $p = \sum_\beta mn_\beta(V^1_\beta)^2$, where the sum is over all V^1_β (positive and negative). The right-hand side is just mn times the average value of $(V^1)^2$ for the gas particles, where $n = \sum_\beta n_\beta$ is the total number of particles per unit volume. For a gas in equilibrium, this is the same as the average value of $(V^2)^2$ or $(V^3)^2$, or one-third of the average value of $|\boldsymbol{V}|^2 = (V^1)^2 + (V^2)^2 + (V^3)^2$. We can therefore write $3p = \sum_b m|\boldsymbol{V}_b|^2$, where the sum is over all n particles in a unit volume of the gas; or $3Bp = \sum_b m|\boldsymbol{V}_b|^2$, where the sum is over all N particles in the volume B. In (3.1) we have written $|\boldsymbol{V}_b| = V_b$.

[3] Fields in physics are like words in human relationships or money in commerce. They begin as convenient labels or counters, but soon gain autonomy and aggressively expand their realms.

[4] Imagine, for example, that the already assembled matter consists of particles fixed in the ϕ_0 frame. The empirical energy of each particle is constant, and the Newtonian energy changes because of the changing potential.

V The Postnewtonian Theory

1. Limitations of gravitostatics

Gravitostatics, strictly interpreted, has a limited range which we have occasionally overshot; some of our more interesting results have been proved in a rather shady manner. We have dealt with dynamical problems by arguing that small, rapidly moving objects produce no appreciable gravitational field, and that large, slowly moving objects produce the same field as when they are stationary. We calculated the relativistic perihelion advance, for example, by neglecting the gravitational field of the planet and assuming that the Sun moves slowly. The weakness of such an *ad hoc* extension of gravitostatics is that there is no quantitative measure of its accuracy. We must now attempt a more honest treatment of dynamics, in which the equations can be expanded in powers of V/c, where V is the speed of the source. Accuracy can then be measured in terms of the lowest powers of V/c that are neglected.

There are other parameters that must be considered, in addition to V/c. For a particle in a bounded orbit, $\psi = \phi - \phi_0$ is of the same order as $(V/c)^2$, and there are also dimensionless parameters proportional to the pressure and internal energy density of the sources. The generalized theory – which is called the *postnewtonian theory* – involves expansions in powers of all of them. It is clumsy but safe. Rather than trying to leap the great gulf between gravitostatics and an exact, relativistic theory, we go step by step, like a spelunker feeling his way up the dark wall of a cave. After we reach the top, we can perhaps retell the story in a neat, axiomatic manner.

We have used a subscript N to denote Newtonian quantities – usually quantities in a ϕ_0 frame. In postnewtonian theories, however, we must introduce more general charts, and the idea of a Newtonian quantity loses significance. We still have to distinguish between mathematical and empirical quantities, but in the course of this chapter we gradually abandon the use of N and E subscripts.

We require a little more special relativity than before. It will be helpful if the reader has met 4 vectors and general Poincaré transformations. Most of the necessary equations are derived in Appendix B.

2. Sources of the gravitational potential

In our development of gravitostatics, we assumed that all forms of energy act in the same way as sources of the gravitational potential. Although this is a good approximation in many cases, it is not exactly true. For example, we showed in the last chapter that it may not hold for the gravitational field energy. We therefore cannot adopt it as a foundation for the postnewtonian theory.

The purpose of this section is to derive a more precise expression for the gravitational potential from the assumption that active and passive gravitational masses are equal. This is taken for granted in Newtonian theory, but so important here that we state it formally.

Hypothesis: Active and passive gravitational masses are equal

However, we do *not* assert that these masses are exactly proportional to the total energy of the sources (or to the inertial mass – which has so far been defined only in the Newtonian theory). We showed in the last chapter how kinetic energy and pressure contribute to the passive gravitational mass; the hypothesis implies that they contribute in the same way to the active gravitational mass. We showed that the gravitational field energy does not contribute to the active gravitational mass (at least if the energy density is IV(4.5)), and it follows that neither does it contribute to the passive gravitational mass.

We begin by reconsidering the field equation of gravitostatics. It was assumed in Chapter III that the dimensionless gravitational potential ϕ is a solution of the Poisson equation III(2.7). With the standard parameters III(3.6), the gravitational constant $k = Gc^{-4}$ is the same in empirical and Newtonian units: we have $k_N = k$ by III(2.8), and III(2.7) becomes $\nabla^2\phi = 4\pi k\epsilon$. The function ϵ was interpreted as the energy density of the sources in Newtonian units, and since $[\epsilon] = [\mathcal{M}\mathcal{L}^{-1}\mathcal{T}^{-2}]$, eq.III(4.1) gives $\epsilon = \epsilon_E e^{-2\psi}$, where $\psi = \phi - \phi_0$, and ϵ_E is the energy density in empirical units (the empirical energy per unit empirical volume). Rewriting III(2.7), og

$$\nabla^2\psi = 4\pi k\epsilon_E e^{-2\psi}. \tag{2.1}$$

Eq.(2.1) implies that the active gravitational mass of a system is proportional to the total Newtonian energy of the sources. In the case considered in Newtonian theory, where the energy is rest energy of stationary particles, this means that the active and passive gravitational masses are equal. When a significant fraction of the energy is in other forms, we shall see that eq.(2.1) implies that the masses may be different, contradicting our hypothesis. If (2.1) is to be compatible with the equality of passive and active gravitational mass, the right-hand side must be modified to contain, not only the energy density of the sources, but terms involving the pressure and velocity. As a preliminary, we must examine more carefully how these quantities are to be defined.

We do not consider the most general possible sources of the gravitational field. Instead, we restrict ourselves to the simplest form of matter that can be expected to give a fairly realistic model of astronomical bodies: an ideal fluid. Living as we do on a stony planet, an ideal fluid may seem to be an overly special choice. Solid bodies are however very rare in the universe; even the Earth has a partly molten core, and behaves gravitationally much like a liquid ball.

The sources – consisting of ideal fluid – have a mass density ρ. This is the proper mass per unit proper volume, measured in empirical units (note that we are no longer using a subscript E to denote empirical units). We recall that proper mass is mass measured in a spacetime chart where the matter is instantaneously at rest, and proper volume is spatial volume ("3 volume") measured in such a chart.

Having defined the mass density ρ at each spacetime point in terms of a particular chart, we define it at all spacetime points in every chart by requiring that it be an invariant. This means that if $x(p)$ and $x'(p)$ are the coordinates of the spacetime point p in the charts x and x', respectively, and ρ and ρ' are the density in x and x', then $\rho'(x'(p)) = \rho(x(p))$ for all p in the domains of both charts. As before, we shall usually suppress explicit mention of the spacetime point p, and write simply $\rho'(x') = \rho(x)$, where x' and x are regarded as quadruples of real numbers. Physics books sometimes write $\rho' = \rho$, which is wrong unless the functions are defined on sets of spacetime points, rather than quadruples of real numbers.

There are some difficulties with the definition of mass density because proper mass is not in general a conserved quantity – the proper mass of a system may change from time to time. We shall discuss this later; for the present we assume the existence of a smooth, invariant function ρ.

The energy density of the sources is written as $\zeta = \rho c^2(1 + \Pi)$. This is proper energy density per unit proper volume measured in empirical units. The constant c is the speed of light measured in empirical units (universal constants are always measured in empirical units), and the invariant, dimensionless function Π is the *internal energy parameter*. We may sometimes call it the *internal energy*, even though it is dimensionless.

In Chapter IV, Section 3, we defined the passive gravitational energy of a body to be the magnitude of the gravitational force on it divided by the value of

$|\nabla\phi|$ in its neighbourhood. The passive gravitational mass is c^{-2} times the passive gravitational energy. The passive gravitational energy is in Newtonian units but the passive gravitational mass is not.

The passive gravitational mass of a volume B of gas, with pressure p and total energy $\mathcal{E}$, is proportional to $\mathcal{E}+3Bp$, from IV(3.4). This suggests that if active and passive gravitational mass are to be equal, then the source term ϵ_E of (2.1) must be modified to include a term $3p$.

In the argument that led to IV(3.4), we assumed that the centre of mass of the gas is at rest. By removing this restriction, we can find the contribution of the kinetic energy to the passive gravitational mass.

Eq.III(4.6) implies that the passive gravitational energy of a particle b with mass m_b, speed $V'_b = |\boldsymbol{V}'_b|$, and energy E'_b is $E'_b(1+V'^2_b c_N^{-2})$. We sum over all the particles in a small element of fluid, assume that c_N does not change appreciably over the element, and write $\boldsymbol{V}'_b = \boldsymbol{V} + \boldsymbol{V}''_b$, where $\boldsymbol{V}$ is the average velocity of the particles in the element (defined so that $\sum_b m_b \boldsymbol{V}''_b = 0$). We find that the total passive gravitational energy of the particles is

$$\mathcal{E} + \sum_b m_b V''^2_b + \mathcal{E} V^2 c_N^{-2}, \tag{2.2}$$

where $\mathcal{E} = \sum_b E''_b$, and we have neglected terms of fourth degree in $V''_b c_N^{-1}$. From IV(3.1), the second term in (2.2) is the same as the "pressure" term in IV(3.4), apart from the replacement of V_b by V''_b and m by m_b (we are allowing the particles to have different masses – which we could also have done in IV(3.1)). The third term in (2.2) implies that if the passive and active gravitational energies (or masses) are to equal, then the source term in the field equation must include a term $\zeta V^2 c_N^{-2} \approx \rho V^2$, where we neglect products of $V^2 c_N^{-2}$, ψ, and Π.

We conclude from the last paragraphs that, if passive and active gravitational masses are to be the same, then the source of the gravitational potential should be proportional to $\zeta + 3p + \rho V^2$, rather than to ζ, where ζ is the proper energy per unit proper volume of the sources in empirical units. However, we cannot however simply replace ϵ_E by $\zeta + 3p + \rho V^2$ in (2.1), because ϵ_E is an energy density, not a proper energy density: it is energy per unit volume, not *proper* energy per unit *proper* volume.[1]

One can find the relationship between energy density and proper energy density by the following elementary argument. The fluid in a small box with proper 3 volume B has proper energy ζB. If V is the speed of the fluid in the chart x, and $\gamma = (1 - V^2c^{-2})^{-1/2}$, then the energy of the material in the box in the chart x is approximately $\gamma\zeta B \approx \zeta B(1 + V^2c^{-2}/2)$ by the usual special relativistic formula for the transformation of energy (or by the argument that $\zeta BV^2c^{-2}/2$ is the kinetic energy of the mass ζBc^{-2}). Note that we are neglecting products of V^2c^{-2} and ψ, so that $V^2c_N^{-2} \approx V^2c^{-2}$, etc. From the FitzGerald contraction, the proper volume B corresponds to a volume B/γ in the chart x, and the energy density in x is therefore approximately $\gamma^2\zeta \approx \zeta(1+V^2c^{-2})$. Taking account of the pressure term and the other velocity term, we must therefore replace ϵ_E by $\zeta(1+V^2c^{-2})+3p+\rho V^2$, and (2.1) becomes

$$\begin{aligned}\nabla^2\psi &= 4\pi k\rho c^2(1+\Pi+2V^2c^{-2}+3p/\rho c^2)e^{-2\psi}\\ &= 4\pi k\rho c^2(1+\Pi+2V^2c^{-2}+3p/\rho c^2-2\psi),\end{aligned} \tag{2.3}$$

where we neglect products of Π, V^2c^{-2}, $p/\rho c^2$, and ψ. The term $p/\rho c^2$ is dimensionless, and we interpret p to be an invariant – the *proper pressure* (it is force per unit area measured in empirical units in a frame where the fluid is instantaneously at rest).

In solving (2.3), we assume that the sources have bounded spatial support, i.e. ρ vanishes, at each instant, outside some finite spatial region. Provided that Π, V^2c^{-2}, $p/\rho c^2$, and ψ are small compared with 1, the approximate solution of (2.3) that is bounded at spatial infinity is $\psi = -U$, where

$$U(x) = kc^2\int \rho(x^0,\boldsymbol{y})|\boldsymbol{x}-\boldsymbol{y}|^{-1}\,d^3y, \tag{2.4}$$

$x = (x^0, \boldsymbol{x})$, $\boldsymbol{x} = (x^1, x^2, x^3)$, $\boldsymbol{y} = (y^1, y^2, y^3)$, $d^3y = dy^1dy^2dy^3$, and the integral is over all y^1, y^2, y^3. If one substitutes $\psi = -U$ in the right-hand side of (2.3), one finds as the next approximation

$$\psi(x) = -kc^2\int \rho(1+\Pi+2V^2c^{-2}+3p/\rho c^2+2U)(x^0,\boldsymbol{y})|\boldsymbol{x}-\boldsymbol{y}|^{-1}d^3y. \tag{2.5}$$

If a body B occupies the spatial region $\Omega(B)$ at time $t = x^0/c$, eq.(2.5) suggests that one should define its active gravitational mass to be

$$\mathcal{M}_B(t) = \int_{\Omega(B)} \rho(1+\Pi+2V^2c^{-2}+3p/\rho c^2+2U)(x)\,d^3x. \tag{2.6}$$

However, this is a time-dependent quantity. In Chapter VIII, eq.VIII(4.12), we define a slightly different mass which is time-independent (see also VIII(4.36) and VIII(4.37)).

In deriving (2.5), we expanded in power series and neglected products of small terms in a rather unsystematic fashion. Before we go on to derive the postnewtonian spacetime metric, we must devise some measure of "smallness," and an appropriate notation. This is the subject of the next section.

3. The postnewtonian approximation

We are trying to construct a theory that is valid for small values of the parameters ψ, Vc^{-1}, $p/\rho c^2$, and Π. The parameters are usually independent, in the sense that one cannot be written as a function of the others, but it is possible to get a rough indication of their relative sizes. For example, we recall from the virial theorem[2] of Newtonian mechanics that, for particles in bounded orbits, the gravitational potential energy is of the same order as the kinetic energy – which implies that ψ is of the same order as V^2c^{-2}. Similarly, p/ρ is of the order of the square of the speed of sound in the fluid, or of the square of the speed of the fluid molecules, and we assume that these speeds are of the same order as V. The internal energy per unit mass of the fluid is $c^2\Pi$, which is of the same order as p/ρ.

We introduce the notation O_n to mean a term of the same order of magnitude as $(V/c)^n$. We shall call this an O_n *term* or a *term of order n*. The results of the last paragraph can then be summarized as

$$V/c = O_1, \qquad \psi = O_2, \qquad p/\rho c^2 = O_2, \qquad \Pi = O_2. \tag{3.1}$$

The product of an O_m term and an O_n term is of course an O_{m+n} term. It is mathematically more satisfactory to regard the postnewtonian approximation as being an expansion in powers of c^{-1}, so that an O_n term is proportional to c^{-n} [Chandrasekhar 67].

If f is a dimensionless quantity, $f = O_m$, we need an estimate of the size of its derivative $\partial_i f$ with respect to the spatial coordinate x^i. For arbitrary f, this is of course impossible; but if we restrict f to be a "reasonable" function of the variables that describe the dynamical system, it is plausible to assume that $L\partial_i f = O_m$ for some constant L with the dimensions of length. One would expect L to be of the same order of magnitude as the diameter of the (bounded) sources, or as the distance in which the density ρ changes by a significant amount. Similarly, for second and higher spatial derivatives, we assume that $L^2\partial_j\partial_i f = O_m$, etc. With (2.3) and (3.1), this implies that

$$k\rho c^2 = L^{-2}O_2. \tag{3.2}$$

If t is the time coordinate, we write the partial derivative of f with respect to $x^0 = ct$ as $\partial_0 f$. We assume that $\partial_0 f$ is of the same order as V/c times the spatial derivatives: $L\partial_0 f = O_{m+1}$ if $f = O_m$. Similarly, for higher derivatives we assume that $L^2\partial_0^2 f = O_{m+2}$, etc. These rules hold for quantities measured in or near the sources – the nearby gravitational potential, for example. They would not be valid if f satisfied something like the wave equation $\nabla^2 f - \partial_0^2 f = 0$ – which by analogy with electromagnetism one might expect far from the sources, in the "radiation zone."

When indicating magnitudes, we often omit powers of L and other dimensional quantities. If $\psi = O_2$, we may write $\nabla^2\psi = O_2$ for example. Nevertheless, we must keep L in mind, not only for dimensional correctness, but as a warning that the postnewtonian approximation is valid only in a bounded spatial region.

Returning to (2.3), we have $k\rho c^2 = L^{-2}O_2$ from (3.2), and the remaining terms are $L^{-2}O_4$. The terms that we have neglected are $L^{-2}O_6$, so we write "$+L^{-2}O_6$" on the right-hand side. Similarly, we write "$+O_6$" on the right-hand side of (2.5).[3]

If the Newtonian expression for a quantity is correct through O_m (i.e. up to and including O_m terms), then an expression correct through O_{m+2} is said to be *postnewtonian*, and some people say that an expression correct through O_{m+4} is *postpostnewtonian*. For example, the Newtonian expression for ψ is $\psi = -U = O_2$, the postnewtonian expression is (2.5) (which contains both O_2 and O_4 terms), and a postpostnewtonian expression would contain O_6 terms also. Note that some authors define O_m to be our O_{2m} – they write $O_{5/2}$ when we would write O_5, for example.

4. Mass densities and continuity equations

Before baring the secrets of the postnewtonian metric, we must consider the problem of the conservation of mass, and derive a postnewtonian continuity equation. But first we review mass conservation in classical fluid mechanics and special relativity, and tensor transformation laws for arbitrary changes of chart.

In classical fluid mechanics, the density ρ^* of the fluid integrated over the spatial region Ω at the instant $t = x^0/c$ gives the mass m of fluid in Ω at t. If Ω is a fixed region (i.e. independent of t), conservation of mass implies that the rate of change of M is equal to the mass flux into Ω across its bounding surface $\partial\Omega$:

$$\begin{aligned} c(d/dx^0)\int_\Omega \rho^*(x^0,\boldsymbol{x})\,d^3x &= \int_\Omega \partial_t\rho^*(x^0,\boldsymbol{x})\,d^3x \\ &= -\int_{\partial\Omega} \rho^* V^m n^m(x^0,\boldsymbol{x})\,dA \\ &= -\int_\Omega \operatorname{div}(\rho^*\boldsymbol{V})(x^0,\boldsymbol{x})\,d^3x = 0, \end{aligned} \tag{4.1}$$

where V^m are the velocity components of the fluid, and n^m are the components of the unit outward normal of $\partial\Omega$. In the last equation of (4.1), we have used the divergence theorem to transform the surface integral into a volume integral. Since Ω is arbitrary, and ρ^* and V^m are assumed to be differentiable functions (at least C^1), it follows that

$$\partial_t\rho^* + \operatorname{div}(\rho^*\boldsymbol{V}) = 0. \tag{4.2}$$

This is the *(nonrelativistic) continuity equation.*

A mere change of notation converts the nonrelativistic continuity equation into the corresponding special relativistic equation. We define $\rho = \rho^*/\gamma$, where $\gamma = (1 - |\boldsymbol{V}|^2 c^{-2})^{-1/2}$, $v^0 = \gamma c$, and $v^i = \gamma V^i$ (or $\boldsymbol{v} = \gamma \boldsymbol{V}$). Eq.(4.2) is then equivalent to $\partial_t(\rho v^0/c) + \text{div}\,(\rho \boldsymbol{v}) = 0$. If we define ∂_0 to be the partial derivative with respect to $x^0 = ct$, eq.(4.2) becomes

$$\sum_\mu \partial_\mu(\rho v^\mu) = 0. \tag{4.3}$$

The summation in (4.3) is over the range of μ, which is $\{0, 1, 2, 3\}$. We remind the reader that lower-case Greek indices (i.e. subscripts or superscripts) always have this range, and lower-case Latin indices have the range $\{1, 2, 3\}$, unless stated otherwise. We also adopt the *summation convention* for lower-case Greek and Latin indices. This says that if an index is repeated in a term, then it is to be summed over its range, unless stated otherwise. By a *term* one means an algebraic expression whose parts are related by multiplications, rather than by additions or subtractions. For example, $a_\mu b^\mu = \sum_\mu a_\mu b^\mu$, but $a_\mu + b_\mu$ has no implied summation.

In (4.3), it is understood that the x^μ are coordinates in an inertial chart x – one of the preferred charts of special relativity. The v^μ are the components in x of the 4 velocity field v of the fluid. It follows from the definition of γ that they satisfy (cf. III(8.4))

$$\eta_{\mu\nu} v^\mu v^\nu = -c^2, \tag{4.4}$$

where $\eta_{mn} = \delta_{mn}$ and $\eta_{\mu 0} = \eta_{0\mu} = -\delta_{\mu 0}$. The $\eta_{\mu\nu}$ are the components in x of the *flat spacetime metric* (or simply of the *metric*, if this is unambiguous).

Transformation laws

Any inertial chart x' is related to the inertial chart x by a *Poincaré transformation:*

$$x'^\mu = L^\mu_\nu x^\nu + l^\mu, \tag{4.5}$$

where the L^μ_ν and l^μ are constants, and (writing $\eta^{\mu\nu} = \eta_{\mu\nu}$ for the sake of elegance)

$$\eta_{\mu\nu} L^\mu_\pi L^\nu_\rho = \eta_{\pi\rho}, \qquad \eta^{\mu\nu} L^\pi_\mu L^\rho_\nu = \eta^{\pi\rho}. \tag{4.6}$$

The first of these equations implies the second (more details in Appendix B). The components of v in x' are $v'^\mu = L^\mu_\nu v^\nu$. They satisfy an equation exactly like (4.4): $\eta_{\mu\nu} v'^\mu v'^\nu = -c^2$.

Exercise 4.1 If f is a Poincaré invariant function of the spacetime coordinates, and ∂_μ denotes the partial derivative with respect to the x^μ coordinate, show that $\eta^{\mu\nu}\partial_\mu\partial_\nu f$ is a Poincaré invariant function. (The operator $\eta^{\mu\nu}\partial_\mu\partial_\nu$ is the *wave* *operator* in the inertial chart.) [Define f' by $f'(x') = f(x)$ for all x, where $x = (x^0, x^1, x^2, x^3)$, $x' = (x'^0, x'^1, x'^2, x'^3)$, and the x'^μ are related to the x^μ by (4.5). Show that $\eta^{\mu\nu}\partial'_\mu\partial'_\nu f'(x') = \eta^{\mu\nu}\partial_\mu\partial_\nu f(x)$. Assume, as always, that all necessary derivatives exist.] ■

It is often convenient to use inertial charts in special relativity, but one is not compelled to do so. Since $L^\mu_\nu = \partial x'^\mu/\partial x^\nu$, the transformation law for the components of v between inertial charts can be written $v'^\mu = (\partial x'^\mu/\partial x^\nu)v^\mu$, and one can regard this equation as the definition of the v'^μ even when x' is not inertial.

Exercise 4.2 If y and y' are any charts,[4] and v^μ and v'^μ are the components of the 4 velocity v in y and y', respectively, show that

$$v'^\mu = (\partial y'^\mu/\partial y^\nu)v^\mu. \tag{4.7}$$

[The v^μ and v'^μ are defined in terms of the components v^μ_I in an inertial chart. You must show that they do not depend on the choice of inertial chart, and that (4.7) holds.] ■

The components $g_{\mu\nu}$ of the (flat spacetime) metric in any chart y are given in terms of the components $\eta_{\mu\nu}$ in the inertial chart x by $g_{\mu\nu} = (\partial x^\pi/\partial y^\mu)(\partial x^\rho/\partial y^\nu)\eta_{\pi\rho}$. One shows (cf. the last exercise) that for any charts y and y' the components of the spacetime metric are related by

$$g'_{\mu\nu} = (\partial y^\pi/\partial y'^\mu)(\partial y^\rho/\partial y'^\nu)g_{\pi\rho}, \tag{4.8}$$

which is equivalent to II(8.9). Note the different placement of primed and unprimed coordinates in (4.7) and (4.8).

If we define functions $v_\mu = g_{\mu\nu}v^\nu$ and $v'_\mu = g'_{\mu\nu}v'^\nu$, and use the fact that $(\partial y'^\mu/\partial y^\nu)(\partial y^\nu/\partial y'^\pi) = \delta^\mu_\pi$, we find that

$$v'_\mu = (\partial y^\nu/\partial y'^\mu)v_\nu. \tag{4.9}$$

(Again for elegance, we have written the Kronecker delta as δ^μ_π rather than $\delta_{\mu\pi}$.)

The transformation laws (4.7)–(4.9) characterize different kinds of *tensor field*. Eq.(4.7) is the transformation law of a 1 contravariant tensor field, or contravariant vector field, or simply vector field (one says *4 vector field* only when the dimension is in doubt); eq.(4.9) is that of a 1 covariant tensor field, or covariant vector field, or covector field; eq.(4.8) is that of a 2 covariant tensor field. For brevity, one often says *tensor* for *tensor field*, *vector* for *vector field*, etc.

Exercise 4.3 If f is an invariant function, show that $\partial_\mu f$ are the components of a covariant vector field. If $g_{\mu\nu}$ are the components of a 2 covariant vector field and $\det(g_{\mu\nu}) \neq 0$ (i.e. the determinant of the matrix with elements $g_{\mu\nu}$ does not vanish), define functions $g^{\mu\nu}$ by $g^{\mu\nu} g_{\nu\pi} = \delta^\mu_\pi$, and show that they are the components of a 2 contravariant tensor field (you must define what that means). If the $g_{\mu\nu}$ are symmetric ($g_{\mu\nu} = g_{\nu\mu}$) show that the $g^{\mu\nu}$ are symmetric ($g^{\mu\nu} = g^{\nu\mu}$). ■

Exercise 4.4 If a^μ and b^μ are components of vector fields, $a_\mu = g_{\mu\nu} a^\nu, b_\mu = g_{\mu\nu} b^\nu$, and $g_{\mu\nu}$ and $g^{\mu\nu}$ are as in the last exercise, show that $g_{\mu\nu} a^\mu b^\nu = g^{\mu\nu} a_\mu b_\nu$, and that this quantity is an invariant. ■

It follows from Exercise 4.4 and (4.4) that in any chart y the 4 velocity v satisfies

$$g_{\mu\nu} v^\mu v^\nu = -c^2. \tag{4.10}$$

Conservation of mass

Assuming that ρ is an invariant (the proper mass per unit proper volume) and that the v^μ are the components of a vector field, one can write the continuity equation (4.3) in terms of an arbitrary chart y. However, it no longer has the same simple form – there are additional terms involving the partial derivatives $\partial y^\mu / \partial x^\nu$. To make the equation *manifestly* covariant (i.e. to make it look the same in all charts), one writes it as

$$\partial_\mu (\Delta \rho v^\mu) = 0, \tag{4.11}$$

where $\Delta = |\det(g_{\mu\nu})|^{1/2}$, and where we reinterpret v^μ and $g_{\mu\nu}$ as components in y, and ∂_μ as being the partial derivative with respect to y^μ. The proof of (4.11) is given in Appendix D. If the reader knows some tensor calculus, she will recall that $\Delta^{-1} \partial_\mu (\Delta \rho v^\mu) = (\rho v^\mu)_{;\mu}$, where the semicolon denotes the covariant derivative (cf. eq.(D14)). Eq.(4.11) is therefore equivalent to the statement that the covariant divergence of the vector field ρv vanishes. Those ignorant of tensor calculus will find (4.11) perfectly adequate.

One defines the components V^m of the *3 velocity* $\mathbf{V}$ in the chart y in terms of the 4 velocity v in y by $V^m / c = v^m / v^0$ (this is consistent with the previous definition for an inertial chart). The *conserved mass density* ρ^* in y is defined by

$$\rho^* = \rho c^{-1} \Delta v^0, \tag{4.12}$$

and (4.11) becomes

$$c \partial_0 \rho^* + \partial_m (\rho^* V^m) = 0. \tag{4.13}$$

Eq.(4.13) has the same form as the continuity equation (4.2) of classical fluid dynamics (apart from the trivial replacement of ∂_t by $c\partial_0$). Provided that the support of the matter distribution is spatially bounded, we show by applying the divergence theorem to (4.13) that $M^* = \int \rho^*(y^0, \boldsymbol{y})\, d^3y$ is independent of y^0. (The integral is over the entire mass distribution, and the domain of the chart y is assumed to be sufficiently large for this to be possible.)

One should remark that ρ^* and M^* are not invariants. They are defined in terms of a chart y, and in general they differ in different charts. We shall formulate the theory in terms of the invariant ρ. Some calculations can however be simplified by using ρ^*.

We have identified the invariant ρ as the proper mass per unit proper volume. If we identify ρ^* as the proper mass per unit coordinate volume, then M^*, which is a conserved quantity, is the total proper mass. This is in agreement with observation for many systems – but not for all. At high energies, proper mass can be created or destroyed (pair production, etc.). More speculatively, there are cosmological theories in which mass is continually being created throughout the universe. We must question whether our interpretations of ρ and ρ^* are exactly correct or only approximations.

One way of dealing with the problem is to assert that baryon number is a strictly conserved quantity, as far as we know, and that in normal matter the ratio of the numbers of baryons and fermions is more or less constant. One can therefore define the mass density ρ to be proportional to the number of baryons per unit proper volume. An objection to this definition is that one can conceive of situations (neutron stars, electron clouds, etc.) where the ratio of the numbers of baryons and fermions differs from that in normal matter. In addition, one cannot be quite certain that baryon number is always conserved – we do not know what happens in strong, time dependent gravitational fields, for example.

We are going to take the point of view that ρ is an invariant that is *defined* to be a solution of (4.11) (and hence that ρ^* is a solution of (4.13)). In a region where we judge that mass in the usual sense is not being created or destroyed, we can identify ρ with the conventional proper mass density (i.e. proper mass per unit proper 3 volume). If the fluid moves through such a region, we identify the mass density there with the conventional mass density, and so find boundary conditions for the solution of (4.11). When this is not possible, we must fix the boundary conditions in some other way – in terms of the proper masses of the particles present at a particular instant, for example. In practice, difficulties rarely arise, and we shall discuss the problem no further.

5. The metric and mass conservation

From now on, we shall usually speak of ϕ_0 *charts* rather than ϕ_0 *frames*. That is, we think of a ϕ_0 chart as being a spacetime chart, like an inertial chart in special relativity, rather than as a purely spatial chart. (In the terminology of Section II.2, a frame consists of a spatial chart together with a time coordinate.)

We showed in Chapter II that the components of the spacetime metric in a ϕ_0 chart have the approximate form

$$g_{00} = -e^{2\psi}, \qquad g_{m0} = g_{0m} = 0, \qquad g_{mn} = \delta_{mn} e^{-2\psi}, \tag{5.1}$$

where we have put $\tau = 1$ and $\lambda = -1$ in II(8.2) and II(8.7). We must now ask how good the approximation is, in postnewtonian terms.

In Section 2 we found an improved expression for ψ, eq.(2.5), which is correct through O_4 (the neglected terms are O_6). Substituting this into the first equation of (5.1), and noting that $\psi = -U + O_4$ with U given by (2.4), og $g_{00} = -e^{2\psi} = -1 - 2\psi - 2\psi^2 + O_6 = -1 - 2\psi - 2U^2 + O_6$. We apply our standard conservative principle – *change things as little as possible* – and assume that there are no other O_4 terms in g_{00}.

The postnewtonian interpretation of the second equation in (5.1) is that the lowest-order terms of g_{m0} vanish. The lowest-order terms are O_1, and we therefore conclude that $g_{m0} = g_m + O_5$, where the g_m are as yet unknown O_3 functions.

In the g_{mn} terms of (5.1), one has only to write $\psi = -U + O_4$. To see why this is sufficient, we consider the invariant $c^{-2} g_{\mu\nu} v^\mu v^\nu$, where v is the 4 velocity. From (4.10), (5.1), and the definition $V^m/c = v^m/v^0$, we have $c^{-1}v^0 = O_0$ (i.e. it is of the same order as the number 1), and $c^{-1}v^m = O_1$. Hence to calculate the invariant through O_4, we have to know g_{00} through O_4, g_{m0} through O_3, and g_{mn} through O_2. We can therefore write $g_{mn} = \delta_{mn} e^{-2\psi} = \delta_{mn}(1+2U) + O_4$. The error term is O_4 rather than O_3 by the argument of invariance under time reversal that we used earlier (see Section 3 and Note 3 at the end of this chapter). The same argument tells us that g_{m0} can be expanded in terms of odd postnewtonian order, as assumed above.[5]

In summary, we assume that the components of the spacetime metric in a ϕ_0 chart x are

$$\begin{aligned} g_{00} &= -e^{2\psi} = -1 + 2U + g_0 + O_6, \\ g_{m0} &= g_m + O_5, \\ g_{mn} &= \delta_{mn} e^{-2\psi} = \delta_{mn}(1 + 2U) + O_4, \end{aligned} \tag{5.2}$$

where ψ is given by (2.5), g_0 consists of the O_4 terms in $-2\psi - 2U^2$, and the g_m are O_3 functions that must be determined.

As a first application of (5.2), we write the postnewtonian form of the continuity equation (4.11). Og $\Delta = |\det{(g_{\mu\nu})}|^{1/2} = e^{-2\psi} + O_6$. The components of the

4 velocity satisfy $g_{\mu\nu}v^\mu v^\nu = -c^2$, eq.(4.10), and are related to the components V^m of the 3 velocity by $V^m/c = v^m/v^0$, so that

$$(v^0/c)^2(g_{00} + 2g_{m0}V^m/c + g_{mn}V^mV^n/c^2) = -1. \tag{5.3}$$

Substituting from (5.2), og $(\Delta v^0/c)^2(-e^{2\psi} + 2g_m V^m/c + e^{2\psi}V^mV^m/c^2 + O_6) = -1$. Expanding the last equation, we find that

$$\Delta v^0/c = 1 + 3U + V^mV^m/2c^2 + O_4, \tag{5.4}$$

and (4.12) and (4.13) give

$$c\partial_0\left[\rho(1 + 3U + V^mV^m/2c^2)\right] + \partial_r\left[\rho V^r(1 + 3U + V^mV^m/2c^2)\right] = \rho c L^{-1}O_5. \tag{5.5}$$

(Recall that a partial time derivative increases the postnewtonian order by 1.) The virtue of using the conserved density ρ^* rather than ρ is that (4.13) is exact, and one does not need clumsy expansions like (5.5).

Exercise 5.1 If $g^{\mu\nu}$ are defined, as in Exercise 4.3, by $g^{\mu\nu}g_{\nu\pi} = \delta^\mu_\pi$, and $g_{\mu\nu}$ are given by (5.2), show that

$$g^{00} = -1 - 2U - g_0 - 4U^2 + O_6, \qquad g^{m0} = g_m + O_5, \qquad g^{mn} = \delta_{mn}(1 - 2U) + O_4. \ \blacksquare \tag{5.6}$$

Exercise 5.2 Show that

$$\begin{aligned} v^0/c &= 1 + U + \Upsilon + O_4, \qquad & v^m/c &= (V^m/c)(1 + U + \Upsilon) + O_5, \\ v_0/c &= -1 + U - \Upsilon + O_4, \qquad & v_m/c &= (V^m/c)(1 + 3U + \Upsilon) + g_m + O_5, \end{aligned} \tag{5.7}$$

where $\Upsilon = V^mV^m/2c^2$. ■

Exercise 5.3 Calculate the O_4 terms in the expressions for v_0/c and v^0/c. ■

6. Motion of particles

In Chapter III, Section 8 we showed that the path of a particle in the metric (5.1) is a geodesic. We have as yet no reason to suppose that the same is true for a particle in the metric (5.2). We shall however calculate its equation of motion on that assumption, so that we can later compare it with the equation of motion found by other means (cf. Chapter VIII, Exercise 2.3).

Eqs.III(8.2)–III(8.5) are still valid, while III(8.6) becomes

$$D\left[e^{-2\psi(z)}v^m + g_m(z)v^0\right]$$

$$= -\left[e^{-2\psi(z)}v^p v^p + e^{2\psi(z)}(v^0)^2\right]\partial_m\psi(z) + \partial_m g_p(z)v^p v^0. \quad (6.1)$$

From III(8.5) og $v^0 = c\gamma e^{-\psi}$, where now

$$\gamma = \left[1 - 2g_m(z)e^{-2\psi(z)}V^m c^{-1} - e^{-4\psi(z)}|\boldsymbol{V}|^2 c^{-2}\right]^{-1/2}, \quad (6.2)$$

and $v^m(\tau) = (dt/d\tau)V^m(t) = v^0(\tau)V^m(t)c^{-1}$. Substituting in (6.1), and for clarity writing $D = \gamma e^{-\psi} d/dt$, og a generalization of III(8.1):

$$(d/dt)(e^{-3\psi}\gamma V^m + g_m c\gamma e^{-\psi}) = -c^2 e^{\psi}\gamma\left[1 + e^{-4\psi}|\boldsymbol{V}|^2 c^{-2}\right]\partial_m\psi$$

$$+ ce^{-\psi}\gamma V^p \partial_m g_p, \quad (6.3)$$

where for brevity we have written ψ for $\psi(\boldsymbol{z},\ .)$ and g_m for $g_m(\boldsymbol{z},\ .)$.

Exercise 6.1 Discuss the effect of the g_r terms in (6.3) on the motion of a planet. [Find the approximate form of the perturbing force on the Kepler motion, and use perturbation theory as in Appendix F.] ■

Notes

[1] To be tediously explicit once more, ϵ_E is empirical energy per unit empirical volume. From III(4.1), $e^{-3\psi}\delta x^1\delta x^2\delta x^3$ is the empirical 3 volume corresponding to the Newtonian or "coordinate" volume $\delta x^1\delta x^2\delta x^3$, and $\epsilon_E e^{-3\psi}\delta x^1\delta x^2\delta x^3$ is the empirical energy in $\delta x^1\delta x^2\delta x^3$. The Newtonian energy in $\delta x^1\delta x^2\delta x^3$ is $\epsilon_E e^{-2\psi}\delta x^1\delta x^2\delta x^3$, from III(4.3), and hence the integral of $\epsilon_E e^{-2\psi}$ with respect to the spatial coordinates x^1, x^2, x^3 gives the total Newtonian energy of the sources. Note that, in this approximation, it does not matter whether the small terms p and ρV^2 are expressed in empirical or Newtonian units. The distinction between proper energy density (measured in a comoving or instantaneous rest frame of the gas) and energy density is of course independent of the choice of empirical or Newtonian units: we can express the proper energy density and the energy density in whichever units we please.

[2] We prove the simplest form of the virial theorem, which is omitted from many elementary texts on Newtonian mechanics. A particle with mass m, displacement $\boldsymbol{Z}$, and velocity $\boldsymbol{V}$ is acted on by a force $\boldsymbol{F}$. We have $\boldsymbol{V} = D\boldsymbol{Z}$ and $mD\boldsymbol{V} = \boldsymbol{F}$. Taking the dot product of the second equation with $\boldsymbol{Z}$, og $(Dm\boldsymbol{V}\cdot\boldsymbol{Z}) - 2T = \boldsymbol{F}\cdot\boldsymbol{Z}$, where $2T = m\boldsymbol{V}\cdot\boldsymbol{V}$ is twice the kinetic energy. For any integrable function f of t, we define the *time average* over the interval $(0, K)$ to be $\langle f\rangle = K^{-1}\int_0^K f(t)\,dt$. We assume that the particle moves in a bounded region of space, and hence that $m\boldsymbol{V}\cdot\boldsymbol{Z}$ is bounded. The time average $\langle m\boldsymbol{V}\cdot\boldsymbol{Z}\rangle$ is then negligible when K is large, and og $2\langle T\rangle = -\langle\boldsymbol{F}\cdot\boldsymbol{Z}\rangle$. In the special case when $\boldsymbol{F} = -\nabla U(\boldsymbol{Z})$ and $U(\boldsymbol{Z}) = b|\boldsymbol{Z}|^{-n}$, where b and n are constants, og $\boldsymbol{F}\cdot\boldsymbol{Z} = nU(\boldsymbol{Z})$, and $\langle T\rangle = -(n/2)\langle U(\boldsymbol{Z})\rangle$. The last equation expresses the virial theorem: the average kinetic energy of a particle in a bounded orbit is $-n/2$ times the average potential energy. For an inverse square law force, one has $n = 1$. The word *virial* has the same root as *virile*, with the sense of *power*, and is a reminder that the theorem applies only to power-law potentials.

[3] Could there be an O_5 term? We can argue not, because (2.3) is invariant under time reversal (the transformation $t \mapsto -t$), and its solution should therefore also be invariant. However, a term with odd powers of V^m/c would change sign. This argument is not completely satisfactory: the O_p term $|\boldsymbol{V}|^p$ does not change sign, for example. We need an additional assumption – one possibility is to require that ψ be an analytic function of the V^m/c (i.e. it can be expressed as a power series in the V^m/c).

[4] We should say any *admissible* charts: there are restrictions on the charts that can be used. If we write the transformation from y to y' as $y'^{\mu} = F^{\mu}(y)$, then the functions F^{μ} must be at least C^2. The same holds for the inverse transformation from y' to y. In addition, the determinant of the Jacobian (the matrix with elements $\partial y'^{\mu}/\partial y^{\nu}$) must not vanish. All these conditions must hold at any point in the domain of definition of both charts. For a proper treatment of such ideas, see the discussion of differentiable manifolds in the differential geometry textbooks.

[5] The argument breaks down if gravitational radiation is emitted – for then time-reversal invariance no longer holds. We assume that this is not significant for the terms of low order considered here.

VI The Postnewtonian Metric

1. Lorentz invariance

In this chapter we follow the same procedure as in the last: we review some results of special relativity, and then modify them very slightly in order to accommodate the gravitational field. The main purpose is to confirm the expressions V(5.2) for the postnewtonian metric, and to determine the functions g_m.

The fundamental hypothesis of special relativity is the Principle of Relativity. It states that any description of a possible set of events in an inertial chart is also the description of a possible set of events in any other inertial chart. (The sets of events are in general different, of course, even though their descriptions are the same.) Another way of putting it is that the set of descriptions of all possible sets of events is the same in all inertial charts – for further discussion, see the end of Appendix B.

Any physical law in special relativity must be compatible with the Principle of Relativity. Sufficient conditions for compatibility are that the law be expressed by the same set of equations in each inertial chart, and that the interpretation of the equations in terms of measurable quantities be the same in each inertial chart. The transformations between inertial charts are called *Poincaré* transformations (or *Lorentz* transformations if the spacetime origin is unchanged – again see Appendix B). A law or theory that is compatible with the Principle of Relativity is said to be Poincaré covariant; if the equations that express it have exactly the same form in each inertial chart, it is said to be *manifestly* Poincaré covariant.

Gravitostatics was formulated in terms of a very restricted set of charts, in which the sources are stationary. The charts are all fixed with respect to one another, and the question of Poincaré covariance is almost trivial. We required only that, in the special case when there is no gravitational field, there is a chart in which the equations of the theory are the same as in an inertial chart. Now, however, we are considering moving sources; we can no longer distinguish a restricted set of charts in which the sources are everywhere at rest; we are forced to consider charts that are moving with respect to one another

We can follow fairly closely the procedure used in special relativity, where the inertial charts are determined as those in which the spacetime metric has a special form ($g_{\mu\nu} = \eta_{\mu\nu}$ in the notation of Section V.4). We look for a set of charts in which the metric is given by V(5.2) with a suitable choice of the functions g_m. In fixing the g_m, we are guided by the assumption that the transformations that relate the charts should be approximately Lorentz transformations, at least for small values of the parameters $V/c, \psi$, etc.

As a preliminary to considering the transformation of the metric, we derive the expression for the potential ψ in a chart x' that is related to the ϕ_0 chart x by a Poincaré transformation. In what follows, we consider Lorentz rather than Poincaré transformations – the shift of origin is trivial. Since rotation of the spatial axes is also fairly trivial, it will be sufficient to restrict ourselves to boosts (which are, roughly speaking, Lorentz transformations without rotation of the spatial axes). It is shown in Appendix B, eq.(B17), that the boost relating the inertial charts x and x' is given by

$$\begin{aligned} x'^0 &= (1 + b^2/2 + 3b^4/8)x^0 - (1 + b^2/2)\boldsymbol{b}.\boldsymbol{x} + LO_5, \\ x'^m &= x^m + b_m \boldsymbol{b} \cdot \boldsymbol{x}/2 - (1 + b^2/2)b_m x^0 + LO_4, \end{aligned} \tag{1.1}$$

whose inverse is

$$\begin{aligned} x^0 &= (1 + b^2/2 + 3b^4/8)x'^0 + (1 + b^2/2)\boldsymbol{b} \cdot \boldsymbol{x}' + LO_5, \\ x^m &= x'^m + b_m \boldsymbol{b} \cdot \boldsymbol{x}'/2 + (1 + b^2/2)b_m x'^0 + LO_4. \end{aligned} \tag{1.2}$$

The boost velocity has been written as $c\boldsymbol{b}$; its m component is cb_m, and its magnitude (the boost speed) is $c|\boldsymbol{b}| = cb$. For inertial charts, $c\boldsymbol{b}$ is the 3 velocity in x of a particle fixed in x' (one sometimes says that $c\boldsymbol{b}$ is the *velocity of x' in x*).

We are going to apply (1.1) and (1.2) to charts x and x' that are not quite inertial, so $c\boldsymbol{b}$ will be not quite the velocity measured in x of a particle fixed in x'. The O_n terms in (1.1) and (1.2) are terms of order b^n, which we identify as terms that are O_n in the postnewtonian sense – i.e. we take the boost speed to be of the same order as the speed V of the sources.

A coordinate transformation that can be written in the form (1.1) will be called a *semiboost*. A coordinate transformation that can be written as the composition of a semiboost with any combination of translations, space reflection, time reversal, and spatial rotations is called a *semipoincaré transformation*. A semipoincaré transformation that leaves the spacetime origin unchanged is a *semilorentz transformation*. In the obvious manner, we speak of a function being semipoincaré invariant, of an equation as being semilorentz covariant, etc.

The potential U is defined in the ϕ_0 chart x by V(2.4), and satisfies the equation $\nabla^2 U = -4\pi kc^2\rho$ (note that this is an exact equation). In the chart x', we define U' by the exactly similar equation

$$U'(x') = kc^2 \int \rho'(x'^0, \boldsymbol{y})|\boldsymbol{x}' - \boldsymbol{y}|^{-1} d^3y. \tag{1.3}$$

It is important to understand that U is *not* an invariant function: in general, $U(x) \neq U'(x')$ when x and x' are related by a Poincaré transformation.

To find the relationship between U and U', it is easiest to introduce a function $\hat{U}$ which *is* an invariant, and which satisfies $\hat{U} = U$ in the chart x. We then have $\hat{U}'(x') = \hat{U}(x) = U(x)$. If the charts x and x' are related by (1.1), og

$$\partial_m = [\delta_{mr} + b_m b_r/2]\partial'_r - b_m \partial'_0 + L^{-1}O_4, \tag{1.4}$$

$$\partial_m \partial_m = \partial'_r \partial'_r + b_r b_s \partial'_r \partial'_s - 2b_r \partial'_r \partial'_0 + L^{-1}O_4, \tag{1.5}$$

and hence, using $\nabla'^2 U' = -4\pi kc^2 \rho'$ and the fact that ρ is an invariant,

$$\begin{aligned} \nabla^2 U(x) &= \nabla'^2 \hat{U}'(x') + b_r b_s \partial'_r \partial'_s \hat{U}'(x') - 2b_r \partial'_r \partial'_0 \hat{U}'(x') + L^{-2}O_6 \\ &= -4\pi kc^2 \rho(x) = -4\pi kc^2 \rho'(x') = \nabla'^2 U'(x'). \end{aligned} \tag{1.6}$$

Solving (1.6), og $U' = \hat{U}' + O_4$, and one can therefore replace $\hat{U}'$ by U' in the O_4 terms of (1.6). It follows from (C42) and (C43) of Appendix C that

$$\begin{aligned} 2\partial'_r \partial'_s U' &= -\nabla'^2 \mathcal{U}'_{rs} + \delta_{rs} \nabla'^2 U', \\ 2\partial'_r \partial'_0 U' &= \nabla'^2 \mathcal{W}'_r - \nabla'^2 \mathcal{V}'_r + L^{-2}O_5, \end{aligned} \tag{1.7}$$

where the potentials $\mathcal{U}'_{rs}$, $\mathcal{V}'_r$, and $\mathcal{W}'_r$ are defined by

$$\mathcal{U}'_{rs}(x') = kc^2 \int \rho'(x'^0, \boldsymbol{y}) K'_r K'_s |\boldsymbol{x}' - \boldsymbol{y}|^{-1}\, d^3y, \tag{1.8}$$

$$\mathcal{V}'_r(x') = kc \int \rho' V'^r(x'^0, \boldsymbol{y})|\boldsymbol{x}' - \boldsymbol{y}|^{-1}\, d^3y, \tag{1.9}$$

$$\mathcal{W}'_r(x') = kc \int \rho' V'^s(x'^0, \boldsymbol{y}) K'_s K'_r |\boldsymbol{x}' - \boldsymbol{y}|^{-1}\, d^3y, \tag{1.10}$$

with $K'_r = (x'^r - y^r)|\boldsymbol{x}' - \boldsymbol{y}|^{-1}$. Substituting in (1.6) and integrating, og $U' = \hat{U}' - (1/2)b_r b_s \mathcal{U}'_{rs} + (1/2)b^2 U' - b_r \mathcal{W}'_r + b_r \mathcal{V}'_r + O_6$, and writing $\hat{U}'(x') = U(x)$ gives

$$U(x) = (1 - b^2/2)U'(x') + (1/2)b_r b_s \mathcal{U}'_{rs}(x') + b_r \mathcal{W}'_r(x') - b_r \mathcal{V}'_r(x') + O_6. \tag{1.11}$$

The potential ψ in the chart x, eq.V(2.5), can be rewritten as

$$\psi = -U - 2\Phi_1 - 2\Phi_2 - \Phi_3 - 3\Phi_4 + O_6, \tag{1.12}$$

where the potentials Φ_1, Φ_2, Φ_3, Φ_4 are defined by

$$\Phi_1(x) = k\int \rho|\boldsymbol{V}|^2(x^0,\boldsymbol{y})|\boldsymbol{x}-\boldsymbol{y}|^{-1}\,d^3y, \tag{1.13}$$

$$\Phi_2(x) = kc^2\int \rho U(x^0,\boldsymbol{y})|\boldsymbol{x}-\boldsymbol{y}|^{-1}\,d^3y, \tag{1.14}$$

$$\Phi_3(x) = kc^2\int \rho \Pi(x^0,\boldsymbol{y})|\boldsymbol{x}-\boldsymbol{y}|^{-1}\,d^3y, \tag{1.15}$$

$$\Phi_4(x) = k\int p(x^0,\boldsymbol{y})|\boldsymbol{x}-\boldsymbol{y}|^{-1}\,d^3y. \tag{1.16}$$

(Recall that $kc^2\rho = L^{-2}O_2$ from V(3.2), and that we should now add a term $L^{-2}O_6$ to V(2.3) and a term O_6 to V(2.5).)

Potentials $\Phi'_\alpha, \alpha \in \{1,\ 2,\ 3,\ 4\}$, are defined in the chart x' by equations exactly similar to (1.13)–(1.16). For example, we write

$$\Phi'_1(x') = k\int \rho' V'^2(x'^0,\boldsymbol{y})|\boldsymbol{x}'-\boldsymbol{y}|^{-1}\,d^3y. \tag{1.17}$$

Like U, the Φ_α are not invariant functions: $\Phi'_\alpha(x') \neq \Phi_\alpha(x)$, in general.

We find the relationship between Φ_α and Φ'_α in the same way that we derived (1.11) for U and U'. Again taking the example of Φ_1, og $\nabla^2\Phi_1 = -4\pi k\rho|\boldsymbol{V}|^2$, $\nabla'^2\Phi'_1 = -4\pi k\rho'|\boldsymbol{V}'|^2$. To lowest order, which is adequate here, $\boldsymbol{V} = \boldsymbol{V}' + c\boldsymbol{b} + cO_3$, and $|\boldsymbol{V}|^2 = |\boldsymbol{V}'|^2 + 2c\boldsymbol{V}'\cdot\boldsymbol{b} + c^2b^2 + c^2O_4$ (ρ, Π, and p are invariants, but $|\boldsymbol{V}|^2$ and U are not). If $\hat{\Phi}_1$ is the invariant function that is defined by $\hat{\Phi}_1 = \Phi_1$ in the chart x, it follows from (1.5) and (1.9) that $\nabla^2\Phi_1(x) = \nabla^2\hat{\Phi}_1(x) = \nabla'^2\hat{\Phi}'_1(x') + L^{-2}O_6 = -4\pi k\rho|\boldsymbol{V}|^2(x) = -4\pi k\rho'|\boldsymbol{V}'|^2(x') - 8\pi kcb_m\rho' V'^m(x') - 4\pi kb^2c^2\rho'(x') + L^{-2}O_6 = \nabla'^2\Phi'_1(x') + 2b_r\nabla'^2\mathcal{V}'_r(x') + b^2\nabla'^2U'(x') + L^{-2}O_6$. Integrating, og

$$\Phi_1(x) = \hat{\Phi}'_1(x') = \Phi'_1(x') + 2b_r\mathcal{V}'_r(x') + b^2U'(x') + O_6. \tag{1.18}$$

Similarly, one shows that

$$\Phi_2(x) = \Phi'_2(x') + O_6, \qquad \Phi_3(x) = \Phi'_3(x') + O_6,$$
$$\Phi_4(x) = \Phi'_4(x') + O_6, \tag{1.19}$$

and that

$$\mathcal{V}_r(x) = \mathcal{V}'_r(x') + b_r U'(x') + O_5,$$
$$\mathcal{W}_r(x) = \mathcal{W}'_r(x') + b_s \mathcal{U}'_{sr}(x') + O_5. \tag{1.20}$$

At last we are ready to write down the transformation law for ψ. We define $\hat{\psi}$ to be an invariant function such that $\hat{\psi}(x) = \hat{\psi}'(x') = \psi(x)$, and we define ψ' by an equation exactly similar to (1.12): $\psi' = -U' - 2\Phi'_1 - 2\Phi'_2 - \Phi'_3 - 3\Phi'_4 + O_6$. Substituting (1.11), (1.18), and (1.19) into (1.12), we find that

$$\begin{aligned}\hat{\psi}' &= -(1 + 3b^2/2)U' - (1/2)b_r b_s \mathcal{U}'_{rs} - b_r \mathcal{W}'_r - 3b_r \mathcal{V}'_r \\ &\quad - 2\Phi'_1 - 2\Phi'_2 - \Phi'_3 - 3\Phi'_4 + O_6 \\ &= \psi' - (3/2)b^2 U' - (1/2)b_r b_s \mathcal{U}'_{rs} - b_r \mathcal{W}'_r - 3b_r \mathcal{V}'_r + O_6. \end{aligned} \tag{1.21}$$

Since $\hat{\psi}'(x') = \psi(x)$, eq.(1.21) gives the transformation law for ψ, and we see that ψ is not an invariant.

We have derived, with some effort, an expression for the gravitational potential in the chart x' that is related to x by the semiboost (1.1). We emphasize that no new physical hypothesis has been made: eq.(1.21) is a consequence of (1.12) and (1.1). It is dispiriting that the expression for ψ' is much more complex than (1.12), and explicitly involves the boost velocity $c\boldsymbol{b}$.

There seems to be little prospect that the metric will have the same form in x and x', and it appears that we may have to abandon the hope of constructing a theory that is approximately Lorentz covariant. Nevertheless, we have come so far that we will continue to the end of the road. Despairing of any happy conclusion, we will bravely transform the spacetime metric. The life of a theoretical physicist is full of frustration, a maze of calculations that lead nowhere.

2. Transformation of the metric

The components $g_{\mu\nu}$ of the spacetime metric in the chart x can be written, from V(5.2) and (1.12), in the form

$$\begin{aligned} g_{00} &= -1 - 2\psi - 2U^2 + O_6 \\ &= -1 + 2U + 4\Phi_1 + 4\Phi_2 + 2\Phi_3 + 6\Phi_4 - 2U^2 + O_6, \\ g_{m0} &= g_m + O_5, \\ g_{mn} &= \delta_{mn}(1 + 2U) + O_4. \end{aligned} \tag{2.1}$$

The potentials U and Φ_α are given explicitly in terms of the sources by V(2.4) and (1.13)–(1.16). All that we know about the g_m is that $g_m = O_3$.

One can easily calculate the components $g'_{\mu\nu}$ of the metric in the chart x' that is related to x by the semiboost (1.1). From V(4.8), the transformation law is $g'_{\mu\nu} = (\partial x^{\pi}/\partial x'^{\mu})(\partial x^{\rho}/\partial x'^{\nu})g_{\pi\rho}$, and og

$$\begin{aligned} g'_{00} &= (1+b^2+b^4)g_{00} + 2b_m g_{m0} + (1+b^2)b_m b_r g_{mr} + O_6, \\ g'_{m0} &= g_{m0} + (1+b^2)b_m g_{00} + (1+b^2/2)b_r g_{rm} + (1/2)b_m b_r b_s g_{rs} + O_5, \\ g'_{mn} &= g_{mn} + b_m b_n g_{00} + b_m g_{n0} + b_n g_{m0} + (1/2)b_m b_r g_{rn} + (1/2)b_n b_r g_{rm} + O_4, \end{aligned} \tag{2.2}$$

where we have omitted arguments x and x' for brevity. Substituting from (2.1), og

$$\begin{aligned} g'_{00} &= -1 - 2\psi - 2U^2 + 4b^2U + 2b_r g_r + O_6, \\ g'_{m0} &= g_m + 4b_m U + O_5, \\ g'_{mn} &= \delta_{mn}(1+2U) + O_4, \end{aligned} \tag{2.3}$$

and (1.11) and (1.21) give

$$\begin{aligned} g'_{00} &= -1 + 2U' - 2U'^2 + 7b^2U' + b_r b_s \mathcal{U}'_{rs} + 2b_r \mathcal{W}'_r + 6b_r \mathcal{V}'_r \\ &\qquad + 4\Phi'_1 + 4\Phi'_2 + 2\Phi'_3 + 6\Phi'_4 + 2b_r g_r + O_6, \\ g'_{m0} &= g_m + 4b_m U' + O_5, \\ g'_{mn} &= \delta_{mn}(1+2U') + O_4. \end{aligned} \tag{2.4}$$

Exercise 2.1 Verify (2.2)–(2.4). ∎

The metric components (2.4) look rather different from (2.1), from which they were obtained by semilorentz transformation. We can make them more similar by choosing g_m in a suitable manner. If we take $2g_m = -7\mathcal{V}_m - \mathcal{W}_m + O_5$, we find from (1.20) that $2b_r g_r = -7b_r\mathcal{V}'_r - 7b^2U' - b_r\mathcal{W}'_r - b_r b_s \mathcal{U}'_{rs} + O_6$. Substituting in (2.4) eliminates the b^2U' and $b_r b_s \mathcal{U}'_{rs}$ terms in g'_{00}:

$$\begin{aligned} g'_{00} &= -1 + 2U' - 2U'^2 + b_r\mathcal{W}'_r - b_r\mathcal{V}'_r + 4\Phi'_1 + 4\Phi'_2 + 2\Phi'_3 + 6\Phi'_4 + O_6, \\ 2g'_{m0} &= -7\mathcal{V}'_m + b_m U' - \mathcal{W}'_m - b_r \mathcal{U}'_{rm} + O_5, \\ g'_{mn} &= \delta_{mn}(1+2U') + O_4. \end{aligned} \tag{2.5}$$

It may seem that (2.5) is not much of an improvement over (2.4): the simplification of g'_{00} is compensated by a complication of g'_{m0}. However, we now show that one can eliminate *all* the $\boldsymbol{b}$ dependent terms from (2.5) by a simple transformation of the time coordinate.

We define a function χ, the *superpotential*, by

$$\chi(x) = -kc^2 \int \rho(x^0, \boldsymbol{y})|\boldsymbol{x} - \boldsymbol{y}|\, d^3y. \tag{2.6}$$

The integral converges because the sources have spatially bounded support: $\rho(x^0, \boldsymbol{x})$ vanishes for large $|\boldsymbol{x}|$. Since $U = O_2$, we have $\chi = L^2 O_2$. Differentiating (2.6), og (cf. Appendix C, eqs.(C6), (C7), (C11))

$$\partial_r \partial_s \chi = -\delta_{rs} U + \mathcal{U}_{rs}, \qquad \nabla^2 \chi = \partial_r \partial_r \chi = -2U, \tag{2.7}$$

$$\partial_0 \partial_r \chi = \mathcal{V}_r - \mathcal{W}_r + O_5. \tag{2.8}$$

We define χ' in the chart x' by an equation exactly like (2.6); it satisfies equations exactly like (2.7) and (2.8). Since $\nabla'^2 \chi' = -2U'$, and $U'(x') = U(x) + O_4$, one proves by an argument similar to the previous ones that $\chi'(x') = \chi(x) + L^2 O_4$. However, a word of warning is necessary: although $\partial'_r \partial'_s \chi'(x') = \partial_r \partial_s \chi(x) + O_4$, it is *not* true that $\partial'_0 \partial'_r \chi'(x') = \partial_0 \partial_r \chi(x) + O_5$ (this would be inconsistent with (1.20)).

If we define $\zeta = (1/2) b_r \partial_r \chi$ and $\zeta' = (1/2) b_r \partial'_r \chi'$, we have $\zeta'(x') = \zeta(x) + LO_5$. Using (2.7) and (2.8), we rewrite (2.5) as

$$\begin{aligned} g'_{00} &= -1 + 2U' - 2U'^2 - 2\partial'_0 \zeta' + 4\Phi'_1 + 4\Phi'_2 + 2\Phi'_3 + 6\Phi'_4 + O_6, \\ 2g'_{m0} &= -7\mathcal{V}'_m - \mathcal{W}'_m - 2\partial'_m \zeta' + O_5, \\ g'_{mn} &= \delta_{mn}(1 + 2U') + O_4. \end{aligned} \tag{2.9}$$

Defining a chart x'' by

$$x''^0 = x'^0 + \zeta'(x'), \qquad x''^m = x'^m, \tag{2.10}$$

og $\partial x''^0 / \partial x'^0 = 1 + \partial'_0 \chi'$, $\partial x''^0 / \partial x'^m = \partial'_m \chi'$, $\partial x''^n / \partial x'^m = \delta_{mn}$, $\partial x''^n / \partial x'^0 = 0$, and the transformation law V(4.8) gives $g'_{00} = g''_{00} - 2\partial'_0 \chi' + O_6$, $g'_{m0} = g''_{m0} - \partial'_m \chi' + O_5$, $g'_{mn} = g''_{mn} + O_4$. Eqs.(2.9) become

$$\begin{aligned} g''_{00} &= -1 + 2U' - 2U'^2 + 4\Phi'_1 + 4\Phi'_2 + 2\Phi'_3 + 6\Phi'_4 + O_6, \\ 2g''_{m0} &= -7\mathcal{V}'_m - \mathcal{W}'_m + O_5, g''_{mn} = \delta_{mn}(1 + 2U') + O_4. \end{aligned} \tag{2.11}$$

Eqs.(2.11) have exactly the same form as (2.1) with $2g_m = -7\mathcal{V}_m - \mathcal{W}_m$. Only one thing remains to be done: we must express the potentials U', Φ'_α, $\mathcal{V}'_m$, and $\mathcal{W}'_m$ of (2.11) in terms of the corresponding potentials U'' defined in the chart x''. Since $\partial'_m = \delta_{mn} \partial''_n + (1/2) b_r \partial'_m \partial'_r \chi' \partial''_0 = \partial''_m + L^{-1} O_4$, og $\nabla'^2 = \nabla''^2 + L^{-2} O_4$, and it follows that $U' = U'' + O_6$, and similarly for the other potentials. One can therefore simply replace U', Φ'_α, $\mathcal{V}'_m$, and $\mathcal{W}'_m$ in (2.11) by the corresponding potentials U'', Φ''_α, $\mathcal{V}''_m$, and $\mathcal{W}''_m$. We define ψ'' by an equation exactly similar to (1.12): $\psi'' = -U'' - 2\Phi''_1 - 2\Phi''_2 - \Phi''_3 - 3\Phi''_4 + O_6$.

In summary, we have assumed that the components of the metric in the chart x are given by (2.1) with $2g_m = -7\mathcal{V}_m - \mathcal{W}_m$:

$$\begin{aligned} g_{00} &= -1 - 2\psi - 2U^2 + O_6 \\ &= -1 + 2U - 2U^2 + 4\Phi_1 + 4\Phi_2 + 2\Phi_3 + 6\Phi_4 + O_6, \\ 2g_{m0} &= 2g_m + O_5 = -7\mathcal{V}_m - \mathcal{W}_m + O_5, \\ g_{mn} &= \delta_{mn}(1 + 2U) + O_4. \end{aligned} \tag{2.12}$$

We have proved that, in the chart x'' defined by

$$\begin{aligned} x''^0 &= (1 + b^2/2 + 3b^4/8)x^0 - (1 + b^2/2)\boldsymbol{b}\cdot\boldsymbol{x} + \zeta(x) + LO_5, \\ x''^m &= x^m + (1/2)b_m\boldsymbol{b}\cdot\boldsymbol{x} - (1 + b^2/2)b_m x^0 + LO_4, \end{aligned} \tag{2.13}$$

the components of the metric have exactly the same form:

$$\begin{aligned} g''_{00} &= -1 - 2\psi'' - 2U''^2 + O_2 \\ &= -1 + 2U'' - 2U''^2 + 4\Phi''_1 + 4\Phi''_2 + 2\Phi''_3 + 6\Phi''_4 + O_6, \\ 2g''_{m0} &= -7\mathcal{V}''_m - \mathcal{W}''_m + O_5, \\ g''_{mn} &= \delta_{mn}(1 + 2U'') + O_4. \end{aligned} \tag{2.14}$$

(Note that in (2.13) we have replaced $\zeta' = (1/2)b_r\partial'_r\chi'$ by $\zeta = (1/2)b_r\partial_r\chi$, which is valid in this approximation.)

Eqs.(2.13) are the same as those for a semiboost, except for the additional term $\zeta(x)$ in the x''^0 equation. The additional term vanishes when $\rho = 0$. We call (2.13) a *postnewtonian boost*, and what we have shown is that the postnewtonian metric is covariant (i.e. unchanged in form) with respect to postnewtonian boosts. A chart in which the metric has the standard postnewtonian form (2.12) is called a *postnewtonian chart*. The transformations that relate postnewtonian charts are *postnewtonian transformations*. They consist of translations, space reflection, time reversal, spatial rotations, and postnewtonian boosts, and any combination of these.

The fact that the metric has the same form in the charts x and x'' means that our original chart x loses its special status. We can now choose any convenient postnewtonian chart with the same freedom as we choose Galilean frames in Newtonian mechanics. The main restrictions are that the sources of the gravitational field be not too massive or fast moving – their speeds in the chart must be small compared with the speed of light. For most astronomical applications of the theory, these are not serious limitations.

Exercise 2.2 Do the postnewtonian transformations form a group? Justify your answer. [Take into account the fact that the speed of the sources must be small compared with the speed of light.] ■

Exercise 2.3 In deriving the postnewtonian transformations, we have assumed a particular form for g_m. Could we have made a more general assumption? [Replace g_m by $g_m + h_m$, and see whether you can remove the additional, unwanted terms from the metric by a coordinate transformation.] ■

The components of the postnewtonian metric (2.12) satisfy the equations

$$2\partial_n g_{mn} - \partial_m g_{nn} + \partial_m g_{00} = L^{-1}O_4, \tag{2.15}$$

$$2\partial_m g_{m0} - \partial_0 g_{mm} = L^{-1}O_5. \tag{2.16}$$

To prove (2.16), note that $\partial_m \mathcal{V}_m - \partial_m \mathcal{W}_m = -2\partial_0 U + L^{-1}O_5$ from (C11) and (C7), and that $\partial_m \mathcal{V}_m = -\partial_0 U + L^{-1}O_5$ from (C12) and the continuity equation V(4.13) or V(5.5). Eqs.(2.15) and (2.16) are called *gauge conditions*. They will be useful in Chapter IX when we consider the postnewtonian approximation of exact field equations.

The metric components (2.12) also satisfy the equations

$$\begin{aligned} \partial_r\partial_r g_{00} &= -4\partial_r U \partial_r U - 8\pi k(\rho c^2 + 2\rho|\mathbf{V}|^2 + \rho c^2 \Pi + 3p) + L^{-2}O_6, \\ \partial_r\partial_r g_{m0} &= -\partial_0\partial_m U + 16\pi k\rho c V^m + L^{-2}O_5, \\ \partial_r\partial_r g_{mn} &= -8\pi k\rho c^2 \delta_{mn} + L^{-2}O_4, \end{aligned} \tag{2.17}$$

where we have used (C33), (C38), (C37), and (C43). We can regard (2.17) as postnewtonian field equations for the $g_{\mu\nu}$. It may seem that field equations are unnecessary, because we already have explicit expressions for the $g_{\mu\nu}$ in terms of the postnewtonian potentials, but they too will be useful in Chapter IX when we consider exact field equations.

3. Postinertial charts

We have derived the postnewtonian metric in charts x, x' and x'', where x and x' are related by the semiboost (1.1), and x' and x'' by the time-coordinate transformation (2.10). The metric has the same functional dependence on the coordinates in x and x'', but not in x' (compare (2.12) and (2.14) with (2.9)). We are now going to define charts $x^{\#}$ and $x'^{\#}$ which are related by a semiboost, and in which the postnewtonian metric has the same functional dependence on the coordinates (the functional dependence is slightly different but no more complicated than in x and x'').

The charts $x^{\#}$ and $x'^{\#}$ are defined by

$$x^{\#0} = x^0 + \theta(x), \qquad x^{\#m} = x^m, \tag{3.1}$$

$$x'^{\#0} = x''^0 + \theta''(x''), \qquad x'^{\#m} = x''^m, \tag{3.2}$$

where

$$\theta(x) = -(1/2)kc \int \rho V^m(x^0, \boldsymbol{y})(x^m - y^m)|\boldsymbol{x} - \boldsymbol{y}|^{-1} d^3y, \tag{3.3}$$

$$\theta''(x'') = -(1/2)kc \int \rho'' V''^m(x''^0, \boldsymbol{y})(x''^m - y^m)|\boldsymbol{x}'' - \boldsymbol{y}|^{-1} d^3y. \tag{3.4}$$

From Appendix C, eq.(C10), og $\theta = -(1/2)\partial_0\chi + LO_5$, and similarly for θ''. Since $V''^m = V^m - cb_m + cO_3$ and $\theta = LO_3$, og $\theta'' = \theta - (1/2)b_m\partial_m\chi + LO_5 = \theta - \zeta + LO_5$. The transformation (3.2) is therefore the composition of (3.1) and the inverse of (2.10). It follows that $x^{\#}$ and $x'^{\#}$ are related by the semiboost (1.1); we shall verify this by an explicit calculation. We need a name to distinguish the new charts $x^{\#}, x'^{\#}$, etc., from the postnewtonian charts. Since they are related by semipoincaré transformations, we shall call them *postinertial charts*.

To calculate the components of the postnewtonian metric in $x^{\#}$, we note that from (C14) and (C11),

$$\partial_0\theta = -(1/2)\partial_0^2\chi + O_6 = (1/2)\Phi_1 + \Phi_4 - (1/2)\mathcal{A} + (1/2)\xi + O_6, \tag{3.5}$$

$$\partial_m\theta = -(1/2)\partial_m\partial_0\chi + O_5 = -(1/2)\mathcal{V}_m + (1/2)\mathcal{W}_m + O_5, \tag{3.6}$$

where the potentials $\mathcal{A}$ and ξ are defined by (C15) and (C16). From (3.1), (3.5), and (3.6) og

$$\begin{aligned} \partial x^{\#0}/\partial x^0 &= 1 + \partial_0\theta + O_6, \qquad \partial x^{\#0}/\partial x^m = \partial_m\theta + O_5, \\ \partial x^{\#m}/\partial x^0 &= 0, \qquad \partial x^{\#m}/\partial x^n = \delta_{mn}. \end{aligned} \tag{3.7}$$

Since $(\partial x^{\#\mu}/\partial x^\nu)(\partial x^\nu/\partial x^{\#\pi}) = \delta^\mu_\pi$, it follows that

$$\begin{aligned} \partial x^0/\partial x^{\#0} &= 1 - \partial_0\theta + O_6, \qquad \partial x^0/\partial x^{\#m} = -\partial_m\theta + O_5, \\ \partial x^m/\partial x^{\#0} &= 0, \qquad \partial x^m/\partial x^{\#n} = \delta_{mn}. \end{aligned} \tag{3.8}$$

The components of the metric in $x^{\#}$ are $g^{\#}_{\mu\nu} = (\partial x^\lambda/\partial x^{\#\mu})(\partial x^\pi/\partial x^{\#\nu})g_{\lambda\pi}$, where the $g_{\lambda\pi}$ are given by (2.12), and og

$$\begin{aligned} g^{\#}_{00} &= g_{00} + 2\partial_0\theta + O_6 = -1 - 2\psi - 2U^2 + 2\partial_0\theta + O_6 \\ &= -1 + 2U + 5\Phi_1 + 4\Phi_2 + 2\Phi_3 + 8\Phi_4 - 2U^2 - \mathcal{A} + \xi + O_6, \\ g^{\#}_{m0} &= g_{m0} + \partial_m\theta + O_5 = -4\mathcal{V}_m + O_5, \\ g^{\#}_{mn} &= \delta_{mn}(1+2U) + O_4. \end{aligned} \tag{3.9}$$

As before (see after (2.11)), we can replace U, Φ_1, etc., on the right-hand sides by $U^{\#}, \Phi_1^{\#}$, etc. Using (3.2) and (2.14), we find exactly similar expressions for the components $g'^{\#}_{\mu\nu}$ of the metric in $x'^{\#}$.

Change of notation For the rest of this section, we drop the # from $x^{\#}$ and $x'^{\#}$. That is, we redefine x and x' to be postinertial charts rather than postnewtonian charts. Similarly, we replace $g^{\#}_{\mu\nu}$ by $g_{\mu\nu}$, and $g'^{\#}_{\mu\nu}$ by $g'_{\mu\nu}$, etc. In later sections, we may use x and x' to denote either postnewtonian or postinertial charts; we shall reintroduce the # symbols only when dealing simultaneously with both kinds of chart. In the new notation, eqs.(3.9) for the components of the metric in the postinertial chart x become

$$\begin{aligned} g_{00} &= -1 + 2U + g^{\#}_0 + O_6, \\ g_{m0} &= g^{\#}_m + O_5, \\ g_{mn} &= \delta_{mn}(1+2U) + O_4, \end{aligned} \tag{3.10}$$

where,

$$\begin{aligned} g^{\#}_0 &= 5\Phi_1 + 4\Phi_2 + 2\Phi_3 + 8\Phi_4 - 2U^2 - \mathcal{A} + \xi, \\ g^{\#}_m &= -4\mathcal{V}_m. \end{aligned} \tag{3.11}$$

It is sometimes convenient to speak of (3.10) as the *postinertial metric* and (2.12) as the *postnewtonian metric* – but they are of course the same metric expressed in different charts. It is more correct to say that (3.10) are the components of the postnewtonian metric in a postinertial chart, and that (2.12) are its components in a postnewtonian chart.

The gauge conditions (2.15) and (2.16) are valid in postinertial charts.

We shall now see how the metric components (3.10) transform under the semiboost (1.1). Since $\partial_0 = (\partial x'^{\mu}/\partial x^0)\partial'_{\mu} = (1 + b^2/2)(\partial'_0 - b_m\partial'_m) + L^{-1}O_5$, and $\chi'(x') = \chi(x) + L^2 O_4$, og

$$
\begin{aligned}
2\partial_0\theta(x) &= -\partial_0^2\chi(x) + O_6 \\
&= (2\partial_0'\theta' + b^2U' - b_m b_r \mathcal{U}'_{mr} + 2b_m\mathcal{V}'_m - 2b_m\mathcal{W}'_m)(x') + O_6, \quad (3.12) \\
2\partial_m\theta(x) &= -\partial_m\partial_0\chi(x) + O_5 \\
&= (2\partial_m'\theta' - b_mU' + b_s\mathcal{U}'_{sm})(x') + O_5. \quad (3.13)
\end{aligned}
$$

The $g_{\mu\nu}$ of (3.10) are the same as those of (2.12), except that a term $2\partial_0\theta$ is added to g_{00} and a term $\partial_m\theta$ to g_{m0}. From (2.4), (3.12), and (3.13) , the additional terms in g'_{00} are $2\partial_0\theta(x)+2b_m\partial_m\theta(x)+O_6 = 2\partial_0'\theta'(x')+O_6$, and in g'_{m0} are $\partial_m\theta(x)+O_5 = (\partial_m'\theta' - (1/2)b_mU' + (1/2)b_s\mathcal{U}'_{sm})(x') + O_5$. Instead of (2.5), we now find

$$
\begin{aligned}
g'_{00} &= -1 + 2U' + 5\Phi_1' + 4\Phi_2' + 2\Phi_3' + 8\Phi_4' - 2U'^2 - \mathcal{A}' + \xi' + O_6, \\
g'_{m0} &= -4\mathcal{V}'_m + O_5, \\
g'_{mn} &= \delta_{mn}(1 + 2U') + O_4,
\end{aligned} \quad (3.14)
$$

which are of the same functional form as (3.9). We have therefore verified that a semiboost preserves the form of the metric in a postinertial chart.

The metric components (3.9) are no more complicated than (2.12): the additional term $\xi - \mathcal{A}$ in the (0, 0) term is compensated by the absence of $\mathcal{W}_m$ in the $(m, 0)$ term. It would be advantageous to adopt (3.9) as the standard form of the postnewtonian metric, and the postinertial charts as the standard charts; the additional transformations (2.10) of the time coordinate would then be unnecessary. However, it is traditional to take (2.12) as the standard form of the metric. We shall use either form, as convenient.

Explicitly covariant form of the postnewtonian metric

We have proved that the functional form of the postnewtonian metric in a postinertial chart is preserved under a semiboost, and more generally under any semipoincaré transformation. We are going to rewrite the metric so that its semipoincaré covariance is obvious (it becomes "manifestly covariant"). The algebra-hating reader can skip this subsection, and Section VII.7 which depends on it, until he reaches Chapter IX.

We first note that the function $\omega_1 = \psi + 2\Phi_1 - \partial_0\theta$ is invariant under the semiboosts (1.1) – to prove this, we have only to substitute from (1.18), (1.21), and (3.12). From (1.12) and (3.5) og $\omega_1 = -U + O_4$ and more precisely

$$
\omega_1 = -U - (1/2)\Phi_1 - 2\Phi_2 - \Phi_3 - 4\Phi_4 + (1/2)\mathcal{A} - (1/2)\xi + O_6. \quad (3.15)
$$

We write the components of the metric in the postinertial chart x in the form

$$g_{\mu\nu} = e^{-2\omega}\eta_{\mu\nu} + \alpha n_\mu n_\nu, \tag{3.16}$$

where α and ω are invariants, n_μ are the components of a covector field, and the $\eta_{\mu\nu}$ are the components of the flat spacetime metric ($\eta_{\mu 0} = -\delta_{\mu 0}$, $\eta_{mn} = \delta_{mn}$). We shall assume that $\alpha = O_0$.

Eq.(3.16) implies the third equation of (3.10) if $\omega = -U + O_4$ and $n_m = O_2$. It implies the second if $\alpha n_m n_0 = -4V_m + O_5$ – which requires that $n_0 = O_1$. To show that it implies the first, we note that $-2\omega_1 = g_0^{\#} + 2U + 2U^2 - 4\Phi_1 + O_6$, from (3.11) and (3.15), and that $e^{-2\omega} = 1 - 2\omega + 2U^2 + O_6$ because $\omega = -U + O_4$. The condition for the first equation to hold is therefore

$$\begin{aligned} 2\omega &= -\alpha n_0^2 - 2\omega_1 + 4\Phi_1 + O_6 \\ &= \alpha\eta_{\mu\nu}n_\mu n_\nu - 2\omega_1 + 4\Phi_1 - \alpha n_m n_m + O_6. \end{aligned} \tag{3.17}$$

The term $\alpha\eta_{\mu\nu}n_\mu n_\nu$ is invariant under semipoincaré transformations. A short calculation using (1.18), (1.20), and (1.11) shows that $\omega_2 = (1/2)\alpha n_m n_m - 2\Phi_1$ is an invariant – and hence that ω is an invariant, as required.

We are free to choose ω in any way compatible with (3.17). An obvious choice is $\omega = \omega_1$. This implies that $\omega = -U + O_4$ and that

$$\alpha n_0^2 = -4\omega_1 + 4\Phi_1 + O_6. \tag{3.18}$$

Another obvious choice is $\omega = \omega_1 + \omega_2$. Again $\omega = -U + O_4$, but now

$$4\omega = \alpha\eta_{\mu\nu}n_\mu n_\nu + O_6. \tag{3.19}$$

More generally, we might write ω as a linear combination of ω_1, ω_2, and any other invariants that we might discover. However, in what follows we shall consider the case $\omega = \omega_1 + \omega_2$ almost exclusively. It allows us to express the metric entirely in terms of the n_μ and α, which will be helpful in Chapter IX when we consider possible generalizations of the postnewtonian field equations.

The $\eta_{\mu\nu}$ are invariant under Poincaré transformations, and α and ω are invariant functions. The metric in the chart x', which is related to x by the semiboost (1.1), is therefore

$$g'_{\mu\nu} = e^{-2\omega'}\eta_{\mu\nu} + \alpha' n'_\mu n'_\nu, \tag{3.20}$$

where $\alpha'(x') = \alpha(x), \omega'(x') = \omega(x)$, and the covector components n'_μ are given in terms of the n_μ by the inverse semiboost (1.2):

$$\begin{aligned} n'_0 &= (1 + b^2/2)n_0 + b_p n_p + O_5, \\ n'_m &= n_m + b_m n_0 + O_4. \end{aligned} \tag{3.21}$$

The O_4 terms $n_m n_n$ do not contribute significantly to the $g_{\mu\nu}$. The transformation law for the $n_\mu n_\nu$ is found from (3.21):

$$\begin{aligned} n_0'^2 &= (1 + b^2)n_0^2 + 2b_m n_0 n_m + O_6, \\ n'_0 n'_m &= n_0 n_m + b_m n_0^2 + O_5, \qquad n'_m n'_n = O_4. \end{aligned} \tag{3.22}$$

We see that the $n'_0 n'_\mu$ are linear combinations of the $n_0 n_\mu$, to the required order. We can therefore describe the postinertial metric in terms of the four variables $n_0 n_\mu$ (and ω, which is related to the n_μ by (3.18) or (3.19)).

If we choose $\omega = \omega_1 + \omega_2$, so that (3.19) is satisfied, we find from (3.15) and the definition $\omega_2 = (1/2)\alpha n_m n_m - 2\Phi_1$ that

$$\begin{aligned} \omega = -U - (5/2)\Phi_1 - 2\Phi_2 - \Phi_3 - 4\Phi_4 + (1/2)\mathcal{A} \\ - (1/2)\xi + (1/2)\alpha n_p n_p + O_6. \end{aligned} \tag{3.23}$$

Og $\partial_0^2 \omega = -\partial_0^2 U + L^{-2} O_6$, where $\partial_0^2 U$ is given by (C36), and from (C33) and (C38) one shows that

$$\begin{aligned} \eta_{\mu\nu}\partial_\mu\partial_\nu\omega = 4\pi k(\rho c^2 + 2\rho|\mathbf{V}|^2 + 2\rho c^2 U + \rho c^2 \Pi + 3p) \\ + (1/2)\partial_m\partial_m(\alpha n_p n_p) + L^{-2}O_6. \end{aligned} \tag{3.24}$$

One finds a similar equation for $\beta = \alpha n_0^2$ by rewriting (3.19) as $\beta = -4\omega + \alpha n_p n_p + O_6$:

$$\begin{aligned} \eta_{\mu\nu}\partial_\mu\partial_\nu\beta = -16\pi k(\rho c^2 + 2\rho|V|^2 + 2\rho c^2 U + \rho c^2 \Pi + 3p) \\ - \partial_m\partial_m(\alpha n_p n_p) + L^{-2}O_6. \end{aligned} \tag{3.25}$$

Similarly, since $\alpha n_m n_0 = -4V_m + O_5$ from (3.10) and (3.16), og

$$\eta_{\mu\nu}\partial_\mu\partial_\nu(\alpha n_0 n_m) = \partial_r\partial_r(\alpha n_0 n_m) + L^{-2}O_5 = 16\pi k c\rho V^m + L^{-2}O_5. \tag{3.26}$$

The field equations can be written in terms of β (or ω) and the variables $q_m = -n_m/n_0$ (this will be useful in Chapter IX). Eq.(3.26) becomes

$$\eta_{\mu\nu}\partial_\mu\partial_\nu(\beta q_m) = \partial_r\partial_r(\beta q_m) + L^{-2}O_5 = -16\pi k c\rho V^m + L^{-2}O_5. \tag{3.27}$$

We have $\alpha n_p n_p = \beta q_p q_p$, and since

$$\begin{aligned}\partial_m(\beta q_p q_p) &= 2\partial_m(\beta q_p)q_p - (\partial_m\beta)q_p q_p, \partial_m\partial_m(\beta q_p q_p)\\ &= 2\partial_m\partial_m(\beta q_p)q_p + 2\beta\partial_m q_p\partial_m q_p - (\partial_m\partial_m\beta)q_p q_p\\ &= -8\partial_m\partial_m(\omega q_p)q_p - 8\omega\partial_m q_p\partial_m q_p\\ &\quad + 4(\partial_m\partial_m\omega)q_p q_p + L^{-2}O_6,\end{aligned} \tag{3.28}$$

og $\eta_{\mu\nu}\partial_\mu\partial_\nu\omega - (1/2)\partial_m\partial_m(\alpha n_p n_p)$

$$= (1-2q_p q_p)\eta_{\mu\nu}\partial_\mu\partial_\nu\omega + 4\partial_m\partial_m(\omega q_p)q_p + 4\omega\partial_m q_p\partial_m q_p + L^{-2}O_6. \tag{3.29}$$

Eq.(3.24) and (3.27) then give

$$\begin{aligned}\eta_{\mu\nu}\partial_\mu\partial_\nu\omega &= 4\pi k(\rho c^2 + 2\rho|\boldsymbol{V}|^2 + 2\rho c^2 U + \rho c^2\Pi + 3p + 2\rho c^2 q_p q_p)\\ &\quad - 16\pi k\rho c V^p q_p - 4\omega\partial_m q_p\partial_m q_p + L^{-2}O_6.\end{aligned} \tag{3.30}$$

Similarly, from (3.25) and (3.28) one shows that

$$\begin{aligned}\eta_{\mu\nu}\partial_\mu\partial_\nu\beta &= -16\pi k(\rho c^2 + 2\rho|\boldsymbol{V}|^2 + 2\rho c^2 U + \rho c^2\Pi + 3p + \rho c^2 q_p q_p)\\ &\quad + 32\pi k\rho c V^p q_p - 2\beta\partial_m q_p\partial_m q_p + L^{-2}O_6.\end{aligned} \tag{3.31}$$

Exercise 3.1 Choose $\omega = \omega_1$, so that n_0 satisfies (3.18), and derive field equations for $\alpha n_0 n_\mu$ and ω. ∎

Outline of solution

Apply the Laplacian operator $\nabla^2 = \partial_m\partial_m$ to (3.18), write $\alpha n_0^2 = \beta$, and use (C33) and (C38) to get

$$\begin{aligned}\partial_m\partial_m\beta &= -4\pi k(4\rho c^2 + 6\rho|\boldsymbol{V}|^2 + 8\rho c^2 U + 4\rho c^2\Pi + 16p)\\ &\quad - 2\partial_m\partial_m\mathcal{A} + 2\partial_m\partial_m\xi + L^{-2}O_6.\end{aligned} \tag{3.32}$$

From (3.18) again, og $\partial_0^2\beta = -4\partial_0^2\omega + L^{-2}O_6 = 4\partial_0^2 U + L^{-2}O_6$, and substituting for $\partial_0^2 U$ from (C36) gives

$$\eta_{\mu\nu}\partial_\mu\partial_\nu\beta = -4\pi k(4\rho c^2 + 4\rho|\boldsymbol{V}|^2 + 8\rho c^2 U + 4\rho c^2\Pi + 12p) + L^{-2}O_6. \tag{3.33}$$

(Note that the $\mathcal{A}$ and ξ terms have disappeared.) Since $\eta_{\mu\nu}\partial_\mu\partial_\nu\omega = -(1/4)\eta_{\mu\nu}\partial_\mu\partial_\nu(\alpha n_0^2) + \partial_m\partial_m\Phi_1 + L^{-2}O_6$ from (3.18), and $\partial_m\partial_m\Phi_1 = -4\pi k\rho|\boldsymbol{V}|^2 + L^{-2}O_6$ from (C38), og

$$\eta_{\mu\nu}\partial_\mu\partial_\nu\omega = 4\pi k(\rho c^2 + 3p + 2\rho c^2 U + \rho c^2\Pi) + L^{-2}O_6. \tag{3.34}$$

In contrast to (3.25) and (3.30), eqs.(3.33) and (3.34) do not have terms non-linear in the field variables. ■

4. Parameterized postnewtonian metric

The postnewtonian metric that we have derived, eq.(2.12), is a special case of the *parameterized postnewtonian metric* (or *PPN metric*). The general form of the PPN metric is written in terms of the potentials U, Φ_α, $\mathcal{U}_{rs}$, $\mathcal{V}_r$, $\mathcal{W}_r$, together with a few more. The potentials are multiplied by constants – the *PPN parameters* – and any observable quantity can be calculated, in the postnewtonian approximation, as a function of the PPN parameters.

Relativistic theories of gravity are usually very complex, and deducing their observational consequences is laborious. The PPN theory was developed to make the calculations easier. One has only to derive the postnewtonian approximation of the relativistic theory, read off the values of the PPN parameters, and substitute them into the known expressions for the observable quantities in the PPN theory. In computer jargon, the PPN theory serves as a universal interface between relativistic theories and observations.

The standard form for the metric in the PPN theory is ([Will 81], eq.(4.48))

$$\begin{aligned} g_{00} &= -1 + 2U - 2\beta U^2 + (2\gamma + 2 + \alpha_3 + \zeta_1 - 2\xi)\Phi_1 \\ &\quad + 2(3\gamma - 2\beta + 1 + \zeta_2 + \xi)\Phi_2 + 2(1+\zeta_3)\Phi_3 \\ &\quad + 2(3\gamma + 3\zeta_4 - 2\xi)\Phi_4 - 2\xi\Phi_W - (\zeta_1 - 2\xi)\mathcal{A} + O_6, \\ 2g_{m0} &= -(4\gamma + 3 + \alpha_1 - \alpha_2 + \zeta_1 - 2\xi)\mathcal{V}_m - (1 + \alpha_2 - \zeta_1 + 2\xi)\mathcal{W}_m + O_5, \\ g_{mn} &= \delta_{mn}(1 + 2\gamma U) + O_4. \end{aligned} \tag{4.1}$$

The PPN parameters are β, γ, ξ, α_1, α_2, α_3, ζ_1, ζ_2, ζ_3, and ζ_4 (the γ here is *not* $(1 - |\boldsymbol{V}|^2 c^{-2})^{-1/2}$). The potential that we have not seen before is

$$\Phi_W(x) = k^2 c^4 \int \rho(x^0, \boldsymbol{y})\rho(x^0, \boldsymbol{y}')K_r(L_r - K_r')|\boldsymbol{x} - \boldsymbol{y}|^{-2}\, d^3y\, d^3y', \tag{4.2}$$

where $K_r = (x^r - y^r)|\boldsymbol{x} - \boldsymbol{y}|^{-1}$, $K_r' = (x^r - y'^r)|\boldsymbol{y} - \boldsymbol{y}'|^{-1}$, $L_r = (y^r - y'^r)|\boldsymbol{x} - \boldsymbol{y}'|^{-1}$. The potentials are chosen to be smooth functions of x that fall off at least as fast as $|\boldsymbol{x}|^{-1}$ when $|\boldsymbol{x}|$ is large. Apart from that, the only real criterion is that they be useful – that they appear in the postnewtonian approximations of the theories that happen to interest us. The scheme is open-ended, in that one can add other potentials as convenient. The potential Φ_W, for example, is called the *Whitehead potential*, and was introduced to handle the Whitehead theory of gravity.

Comparing (2.12) and (4.1), we find that the values of the PPN parameters for our theory are

$$\beta = \gamma = 1, \qquad \xi = \alpha_1 = \alpha_2 = \alpha_3 = \zeta_1 = \zeta_2 = \zeta_3 = \zeta_4 = 0. \tag{4.3}$$

These are usually called the *standard* values, and they seem to be in good agreement with experiment. From a rabidly Newtonian point of view, the greater generality of the PPN theory may appear to be redundant, a source of unnecessary complication. But it would be unwise to be so dogmatic. One can argue that the only proper way to establish that the parameters have the standard values is to analyze all the observational data in the context of the full postnewtonian theory, and show that (4.3) hold with small experimental errors.

VII Postnewtonian Fluid Mechanics

1. Matter in a gravitational field

We have learned how to calculate the spacetime metric due to a given distribution of matter. We assumed that the matter (which may be called the *source* or *sources* of the metric, or of the gravitational field) is a spatially bounded, ideal fluid whose speed, pressure, internal energy density, and gravitational potential are small. We must now solve the complementary problem of how this ideal fluid responds to a given gravitational field.

We determined the motion of particles in gravitostatics by generalizing the special-relativistic Lagrangian of a free particle in an obvious manner. We are going to derive the equations of motion of an ideal fluid in the postnewtonian theory by a similar stratagem. After a brief review of non-relativistic fluid mechanics, we summarize the necessary equations of special relativistic fluid mechanics, and then transfer them, virtually without change, to the gravitational theory.

This chapter is devoted almost entirely to development of the theory; the next shows how the theory is applied in determining the motion of gravitating bodies. We try to give a self-contained, elementary account which can also serve as an introduction to the treatises on the subject [Will 81], [Soffel 89]. To make the transition easier, we sometimes cross-reference our equations to theirs. Other applications of the postnewtonian theory are given in [Misner 73]; readers of Russian should consult [Brumberg 72].

The equations of postnewtonian fluid mechanics are fairly complex, and the reader must be prepared to work through a good deal of algebra in deriving them. Fortunately, one can often obtain useful information without solving the equations explicitly. The conservation laws for energy, momentum, and angular momentum, which are derived in Section 5, are particularly helpful.

2. Nonrelativistic fluid mechanics

Fluid mechanics is often neglected in the physics curriculum. We shall give a brief account of it for the simplest case of an ideal fluid (i.e. we neglect viscosity, etc.). The discussion is a continuation of that in Chapter V, Section 4, where we dealt with mass conservation and the equation of continuity.

We begin with the nonrelativistic theory. We calculate the force on a small piece of fluid, and derive the equations of motion by applying Newtonian mechanics. We first consider an element of area A in the fluid with unit normal $\boldsymbol{n}$. The force $\boldsymbol{F}$ that acts across A has components $F^m = T^{mn}n^n A$, where T^{mn} are the components of the *stress tensor*. (We work in a cartesian chart, so that it does not matter whether tensor indices are written as subscripts or superscripts.) We adopt the convention that the force $\boldsymbol{F}$ acts on the fluid which is on the side of the element to which $\boldsymbol{n}$ points. It then follows from Newton's Third Law that a force $-\boldsymbol{F}$ acts on the fluid which is on the other side.

The fluid has density ρ^* and velocity $\boldsymbol{V}$. The m component of the momentum in a volume Ω at time $t = x^0/c$ is $\int_\Omega \rho^* V^m(x)\, d^3x$, where $(x) = (x^0, \boldsymbol{x})$. The surface of Ω is denoted by $\partial\Omega$. The m component of the force acting on the fluid in Ω is $-\int_{\partial\Omega} T^{mn}n^n(x)\, dA = -\int_\Omega \partial_n T^{mn}(x)\, d^3x$, by the divergence theorem: we recall that, with the usual conventions of the divergence theorem, $\boldsymbol{n}$ is the *outward* pointing normal. We take Ω to be a volume fixed in the cartesian chart, so that the m component of the momentum flux out of Ω is $\int_{\partial\Omega} \rho^* V^m V^n n^n(x)\, dA = \int_\Omega \partial_n(\rho^* V^m V^n)(x)\, d^3x$. In addition to the surface forces, we assume that a force density $\boldsymbol{w}$ acts on the fluid in Ω. Setting the rate of change of momentum equal to the total force, og

$$(d/dt)\int_\Omega \rho^* V^m(x)\, d^3x + \int_\Omega \partial_n(\rho^* V^m V^n)(x)\, d^3x$$
$$= -\int_\Omega \partial_n T^{mn}(x)\, d^3x + \int_\Omega w^m(x)\, d^3x. \tag{2.1}$$

The volume Ω is arbitrary, and ρ^*, V^m, w^m and T^{mn} are assumed to be smooth (C^1 is sufficient). It follows that

$$c\partial_0(\rho^* V^m) + \partial_n(\rho^* V^m V^n) = -\partial_n T^{mn} + w^m. \tag{2.2}$$

This is the equation of motion of the fluid – the non-relativistic Euler equation. For an ideal fluid, the force $\boldsymbol{F}$ across an element of area A is parallel to the normal $\boldsymbol{n}$, and og $T^{mn} = p\delta_{mn}$ and $\partial_n T^{mn} = \partial_m p$, where $p \geq 0$ is the *pressure*. If $\boldsymbol{w}$ is the gravitational force per unit volume and ψ is the dimensionless potential, we write $w^m = -\rho^* c^2 \partial_m \psi = \rho^* c^2 \partial_m U$, where $U = -\psi$ is given by V(2.4), and (2.2) becomes

$$c\partial_0(\rho^* V^m) + \partial_n(\rho^* V^m V^n) = -\partial_m p + \rho^* c^2 \partial_m U. \tag{2.3}$$

We recall that the continuity equation V(4.2) is also satisfied:

$$c\partial_0\rho^* + \partial_n(\rho^* V^n) = 0. \tag{2.4}$$

In this approximation, we do not have to distinguish ρ^* from the invariant density ρ of special relativity or the postnewtonian theory, which we introduced in Chapter V.

If we write $c\partial_0(\rho^* V^m) = c(\partial_0\rho^*)V^m + c\rho^*\partial_0 V^m = -\partial_n(\rho^* V^n)V^m + c\rho^*\partial_0 V^m$ from (2.4), and define $D_t V^m = c\partial_0 V^m + V^r\partial_r V^m$, eq.(2.3) becomes

$$\rho^* D_t V^m = -\partial_m p + \rho^* c^2 \partial_m U. \tag{2.5}$$

We call $D_t V^m$ the m component of the *comoving* or *material* derivative of $\boldsymbol{V}$ (with respect to t).

3. Special-relativistic fluid mechanics

We can derive special relativistic fluid mechanics by the usual device of finding Lorentz covariant equations that reduce to known, nonrelativistic equations when speeds are small compared with the speed of light. Again we try to give brief, self-contained proofs of the essential equations for an ideal fluid.

We first recall some definitions and notation from Chapter V, mainly Section 4. The components of the spacetime metric in an inertial chart are $g_{\mu\nu} = \eta_{\mu\nu}$. The density of a fluid is ρ (the proper mass per unit proper volume) or $\rho^* = \rho\gamma = \rho(1 - |\boldsymbol{V}|^2/c^2)^{-1/2}$ (the proper mass per unit coordinate volume); the former is a Poincaré invariant, the latter is not. The energy density is $\zeta = \rho c^2(1 + \Pi)$ – this is proper energy per unit proper volume, a Poincaré invariant. The 4 velocity of the fluid is a vector field v whose components in an inertial chart satisfy $\eta_{\mu\nu}v^\mu v^\nu = -c^2$, eq.V(4.4). The components of the 3 velocity $\boldsymbol{V}$ of the fluid are given by $V^m/c = v^m/v^0$, and it follows that $v^0 = c\gamma$ (we choose the positive sign).

Given any spacetime point q, one can find an inertial chart x_q with origin at q, in which the 3 velocity of the fluid vanishes at the origin. In terms of the 4 velocity v, this means that $v^m(0) = 0, v^0(0) = c$, where $(0) = (0, 0, 0, 0)$. We call such a chart an *instantaneous rest chart* of the fluid at q.

The components of the *stress-momentum* of the fluid at the spacetime point q are defined in the chart x_q by

$$T^{00}(0) = \zeta(0), \qquad T^{m0} = T^{0m} = 0, \qquad T^{mn}(0) = p(0)\delta_{mn}, \tag{3.1}$$

where ζ and p are the energy density and the pressure. Eqs.(3.1) are equivalent to

$$T^{\mu\nu} = pg^{\mu\nu} + (\zeta + p)v^\mu v^\nu c^{-2} \tag{3.2}$$

at the origin of the chart x_q, where $g^{\mu\nu} = \eta^{\mu\nu} = \eta_{\mu\nu}$. Since ζ and p are invariant, (3.2) is a tensor equation, valid at q in any chart. In addition, the spacetime point q can be chosen arbitrarily, and it follows that (3.2) is valid everywhere and in any chart.

So far we have merely defined the stress-momentum to be the tensor field with the components (3.2). We now assume, as a physical law, that

$$\partial_\nu T^{\mu\nu} = 0, \tag{3.3}$$

provided that no forces act on the fluid except the internal pressure forces. We call $\partial_\nu T^{\mu\nu}$ the *(ordinary) divergence* of the stress-momentum; it behaves like a contravariant vector field under Lorentz transformations.

We interpret T^{00} as the energy density of the fluid in any inertial chart, and T^{m0} as being proportional to the momentum density. Readers familiar with relativistic electromagnetism will recall that the electromagnetic stress-momentum has exactly similar properties – and that the divergence of the electromagnetic stress-momentum vanishes when no charges are present, in analogy with (3.3).

In what follows, we often use the less cumbersome notation $f_{,\mu}$, instead of $\partial_\mu f$, for the partial derivative of f with respect to x^μ. We write (3.3) in more explicit form by substituting from (3.2), multiplying by $g_{\pi\mu} = \eta_{\pi\mu}$, and defining $v_\pi = g_{\pi\mu} v^\mu$:

$$c^2 p_{,\pi} + (\zeta_{,\nu} + p_{,\nu}) v_\pi v^\nu + (\zeta + p)(v_\pi v^\nu)_{,\nu} = 0. \tag{3.4}$$

From V(4.4), og $\eta_{\mu\nu} v^\mu v^\nu = v^\mu v_\mu = \eta^{\mu\nu} v_\mu v_\nu = -c^2$, and hence $v^\mu v_{\mu,\pi} = (1/2)(\eta^{\mu\nu} v_\mu v_\nu)_{,\pi} = 0$. Multiplying (3.4) by v^π and using the last equation, og

$$\zeta_{,\nu} v^\nu + (\zeta + p) v^\nu_{,\nu} = 0, \tag{3.5}$$

and hence

$$c^2 p_{,\pi} + p_{,\nu} v_\pi v^\nu + (\zeta + p) v_{\pi,\nu} v^\nu = 0. \tag{3.6}$$

Eqs.(3.5) and (3.6) are equivalent to (3.3). Eq.(3.6) is the special-relativistic Euler equation for an ideal fluid on which no external forces act (the only forces are the internal pressure forces).

Exercise 3.1 Show that, for small velocities, (3.6) implies (2.3) (with $U = 0$). ■

We have formulated fluid mechanics in terms of velocity fields, etc., that are functions of the spacetime coordinates – this is sometimes called the *Eulerian* formulation. The 4 velocity of the element of fluid that has spatial coordinates $\boldsymbol{x}$ at the time $t = x^0/c$ is $v(x) = v(x^0, \boldsymbol{x})$, but different elements of fluid are at $\boldsymbol{x}$ at different times. In the alternative, *Lagrangian* formulation, one considers the paths (or trajectories) of elements of fluid, in much the same way as in particle dynamics. More precisely, one defines a family of curves $z : \mathbb{I} \to \mathbb{R}^4$ as the solutions of the differential equation $Dz^\mu(\tau) = v^\mu(z(\tau))$, where $\mathbb{I} \subset \mathbb{R}$, $z = (z^0, z^1, z^2, z^3)$, $\tau \in \mathbb{I}$ is a real parameter, and D is the derivative with respect to τ. Each curve z is the path of an element of fluid, and Dz is its 4 velocity. Since $\eta_{\mu\nu} Dz^\mu Dz^\nu = \eta_{\mu\nu} v^\mu(z) v^\nu(z) = -c^2$, we call τ the *proper time* of the fluid element, in analogy with the previous terminology for particles.

For any smooth function f of the spacetime parameters, we define the *comoving* or *material* derivative of f (with respect to τ) by $D_\tau f(z) = (Dz^\mu)\partial_\mu f(z) = v^\mu(z)\partial_\mu f(z)$, or $D_\tau f = v^\mu \partial_\mu$f. We recall that in Section 2 we defined the comoving derivative with respect to t in a similar manner. Note that D acts on a function of one real variable, while D_τ acts on a function of the four spacetime variables. In terms of the material derivative, we can rewrite (3.6) as

$$\zeta D_\tau v_\pi + D_\tau(p v_\pi) = -c^2 p_{,\pi}. \tag{3.7}$$

Special relativistic equations in covariant form

Eqs.(3.2)–(3.7) are valid in any inertial chart. We now write them in covariant form, valid in non-inertial charts also.

Eq.(3.2) is already in covariant form, if we interpret $g^{\mu\nu}, T^{\mu\nu}$, and v^μ as components in an arbitrary chart y. Using eq.D(13) of Appendix D, we show that the covariant form of (3.3) is

$$2(\Delta T^\nu_\mu)_{,\nu} - \Delta g_{\nu\lambda,\mu} T^{\nu\lambda} = 0, \tag{3.8}$$

where $T^\nu_\mu = g_{\mu\lambda} T^{\lambda\nu}, \Delta = |\det(g_{\mu\nu})|^{1/2}$, and the comma denotes the partial derivative with respect to the coordinates y^π. In terms of the covariant derivative, (3.8) is equivalent to $T^\nu_{\mu;\nu} = 0$, from ($D14$). However, we shall not use covariant derivatives in this chapter.

We again write $v_\pi = g_{\pi\mu} v^\mu$, and have $v^\mu v_\mu = g_{\mu\pi} v^\mu v^\pi = -c^2$. The $g_{\mu\pi}$ are in general not constant, and og $g_{\mu\pi}(v^\mu v^\pi)_{,\nu} = -\ g_{\mu\pi,\nu} v^\mu v^\pi$, and

$$v^\mu v_{\mu,\nu} = v^\mu (g_{\mu\pi} v^\pi)_{,\nu} = g_{\mu\pi,\nu} v^\mu v^\pi + (1/2) g_{\mu\pi}(v^\mu v^\pi)_{,\nu} = (1/2) g_{\mu\pi,\nu} v^\mu v^\pi. \tag{3.9}$$

We also need eq.(D9):

$$g_{\lambda\nu,\mu} g^{\lambda\nu} = 2\Delta^{-1}\Delta_{,\mu}. \tag{3.10}$$

Exercise 3.2 Show that

$$\zeta_{,\nu} v^\nu + (\zeta + p)(v^\nu_{,\nu} + \Delta^{-1}\Delta_{,\nu} v^\nu) = 0, \tag{3.11}$$

and hence

$$c^2p_{,\pi} + p_{,\nu}v_\pi v^\nu + (\zeta + p)(v_{\pi,\nu}v^\nu - (1/2)g_{\lambda\nu,\pi}v^\lambda v^\nu) = 0. \qquad (3.12)$$

[Multiply (3.8) by v^μ and use (3.2) and (3.10). To derive (3.12), use (3.11) to eliminate $v^\nu_{,\nu}$.] ■

Exercise 3.3 Show that the four equations in (3.12) are not independent. [Multiply by v^π.] ■

Eqs.(3.11) and (3.12) are equivalent to (3.8) (prove it!); they are the covariant versions of (3.5) and (3.6). By using (3.9), we can write (3.12) in the alternative forms

$$c^2p_{,\pi} + p_{,\nu}v_\pi v^\nu + (\zeta + p)(v_{\pi,\nu} - v_{\nu,\pi})v^\nu = 0, \qquad (3.13)$$

$$c^2p_{,\pi} + D_\tau(pv_\pi) + \zeta D_\tau v_\pi - (\zeta + p)v_{\nu,\pi}v^\nu = 0. \qquad (3.14)$$

Exercise 3.4 Show that, when $p = 0$, eq.(3.12) reduces to the geodesic equation III(8.3). [Write $D(g_{\pi\rho}v^\rho) = Dv_\pi = v_{\pi,\nu}v^\nu$. Note that in Chapter III the v^π are functions of the proper time, while here they depend on the four spacetime variables: it would be better to use different symbols in your proof.] ■

The covariant form of the special relativistic continuity equation is V(4.11): $(\Delta\rho v^\nu)_{,\nu} = 0$. It is equivalent to $\rho\Delta^{-1}\Delta_{,\nu}v^\nu + \rho v^\nu_{,\nu} = -\rho_{,\nu}v^\nu$, and substituting in (3.11)

og $\zeta_{,\nu}v^\nu = (\zeta + p)\rho^{-1}\rho_{,\nu}v^\nu$. We adopt the Lagrangian point of view, write $\zeta_{,\nu}v^\nu = D_\tau\zeta$ and $\rho_{,\nu}v^\nu = D_\tau\rho$, and put $\zeta = \rho c^2(1 + \Pi)$ as before. We find that

$$c^2D_\tau\Pi = -pD_\tau(\rho^{-1}). \qquad (3.15)$$

Since Πc^2 is the internal energy per unit proper mass of the fluid, and ρ^{-1} is the proper volume per unit proper mass, eq.(3.15) can be interpreted as saying that the work done by the pressure forces on the fluid is equal to the increase in the internal energy. Note that there is no entropy term in (3.15) – the stress-momentum (3.2) does not allow dissipation or heat flow, and the entropy stays constant.

4. Postnewtonian fluid mechanics

In Section 3, we dealt with special-relativistic fluid mechanics, which applies only in the absence of a gravitational field. When a gravitational field is present, we might expect the theory to be modified; in fact, we take it to be the same.

More precisely, the equations of motion of an ideal fluid in the presence of a gravitational field are assumed to have the same form as the *generally covariant*, special relativistic equations. This means that, in the special relativistic equations written for an arbitrary chart, one has only to reinterpret $g_{\mu\nu}$ and $g^{\mu\nu}$ as components of the general spacetime metric. We can therefore take the equations of motion of the fluid to be (3.8), or equivalently to be the internal energy equation (3.15) and the Euler equations (3.12) with $\pi \in \{1,2,3\}$ (the $\pi = 0$ equation is a linear combination of the other three, by Exercise 3.3). In addition, the covariant form V(4.11) of the mass continuity equation must be satisfied.

These equations of motion are the ones most commonly postulated in relativistic theories of gravity. They do not depend in any way on the validity of the postnewtonian approximation; the only question is whether they are correct. They are of course the simplest possible generalization of the special relativistic equations, and we shall show that they do imply the standard nonrelativistic equations for the case of weak gravitational fields. Beyond that, one must wait and see whether they give results in agreement with experiment.

We calculate the postnewtonian form of the equations of motion by using the expressions VI(2.12) for the metric in a postnewtonian chart x. For brevity, we write

$$\begin{gathered} g_{00} = -1 + 2U + g_0 + O_6, \\ g_{m0} = g_m + O_5, \qquad g_{mn} = \delta_{mn}(1+2U) + O_4, \end{gathered} \tag{4.1}$$

where $g_0 = 4\Phi_1 + 4\Phi_2 + 2\Phi_3 + 6\Phi_4 - 2U^2 = O_4$ and $g_m = -\,(7\mathcal{V}_m \;+\; \mathcal{W}_m)/2 = O_3$. If the chart x is postinertial, as defined in Section VI.3 then the metric is given by VI(3.10) – which is the same as (4.1), except that the g_μ are replaced by $g^{\#}_\mu$, defined by VI(3.11). From VI(3.9), we have $g^{\#}_0 = g_0 + 2\partial_0\theta + O_6$, $g^{\#}_m = g_m + \partial_m\theta + O_5$, where θ is defined by VI(3.3). In this section and the next we consider only postnewtonian charts. In Section 6 we list the corresponding results for postinertial charts.

We shall need expressions for the components of the 4 velocity more precise than V(5.7), which the virtuous reader has already calculated in Exercise V(5.3):

$$\begin{aligned} v^0c^{-1} &= 1 + U + \Upsilon + 3U^2/2 + 3\Upsilon^2/2 + 5U\Upsilon + g_0/2 + g_mV^mc^{-1} + O_6, \\ v^mc^{-1} &= V^mc^{-1}(1 + U + \Upsilon) + O_5, \\ v_0c^{-1} &= -1 + U - \Upsilon + U^2/2 - 3U\Upsilon - 3\Upsilon^2/2 + g_0/2 + O_6, \\ v_mc^{-1} &= V^mc^{-1}(1 + 3U + \Upsilon) + g_m + O_5, \end{aligned} \tag{4.2}$$

where $\Upsilon = (1/2)V^mV^mc^{-2}$ and we have used the $g^{\mu\nu}$ of V(5.6). If D_τ is the material derivative with respect to the proper time τ, and D_t is the material derivative with respect to the coordinate time t, then $D_\tau = (dt/d\tau)D_t$. Since $v^0 = Dz^0 = c\,dt/d\tau$, og $D_\tau = v^0c^{-1}D_t = (1 + U + \Upsilon + O_4)D_t$, from (4.2).

We take the Euler equations to be (3.14), and calculate their postnewtonian form for $\pi \in \{1,2,3\}$ (as in (3.12), the $\pi = 0$ equation is a linear combination of the other three). We find that

$$\begin{aligned} D_\tau(pv_m) &= D_t(pV^m) + \rho c^4 L^{-1}O_6, \\ D_\tau v_m &= (1+4U+2\Upsilon)D_tV^m + V^m D_t(3U+\Upsilon) + D_t g_m + c^2L^{-1}O_6, \\ c^{-2}v^\nu v_{\nu,m} &= (1+2U+4\Upsilon)U_{,m} + (1/2)g_{0,m} + V^r c^{-1} g_{r,m} + L^{-1}O_6. \end{aligned} \tag{4.3}$$

Substituting (4.3) into (3.14), og to lowest order

$$p_{,m} + \rho D_t V^m - \rho c^2 U_{,m} = \rho c^2 L^{-1} O_4, \tag{4.4}$$

which is equivalent to (2.5). We have therefore established that the general Euler equations do have the correct nonrelativistic limit.

Multiplying (4.4) by V^m gives

$$V^m p_{,m} + \rho c^2 D_t\Upsilon - \rho c^2 V^m U_{,m} = \rho c^3 L^{-1} O_5. \tag{4.5}$$

Since $V^m\partial_m = D_t - c\partial_0$, eq.(4.5) is equivalent to

$$D_t p - cp_{,0} + \rho c^2 D_t\Upsilon - \rho c^2 D_t U + \rho c^3 U_{,0} = \rho c^3 L^{-1} O_5. \tag{4.6}$$

We use (4.6) to eliminate $D_t\Upsilon$ from (4.3), then substitute in (3.14) and find that

$$\begin{aligned} &\zeta(1+2\Upsilon+p/\zeta)D_tV^m + \zeta D_t(4UV^m + cg_m) \\ &\quad - \zeta V^m cU_{,0} + cV^m p_{,0} + c^2 p_{,m} - pc^2 U_{,m} - \zeta c^2(U_{,m} + 2UU_{,m} + 4\Upsilon U_{,m} \\ &\quad + (1/2)g_{0,m} + V^r c^{-1} g_{r,m}) = \rho c^4 L^{-1} O_6, \end{aligned} \tag{4.7}$$

where g_m and g_0 are defined after (4.1).

Eqs.(4.7), together with the continuity equation V(5.5), the internal energy equation (3.15), and the expressions for the spacetime metric (4.1), should enable us to calculate the evolution of any gravitating, spatially bounded, ideal fluid system. The equations are quite complex and we cannot hope for a general solution. As in Newtonian mechanics, we must look for ways of extracting information without explicitly solving them, and for special solutions of physical interest.

When attempting to solve the equations, it is often easier to use the conserved mass density $\rho^* = \rho c^{-1}\Delta v^0$, V(4.12), which satisfies the exact continuity equation $c\rho^*_{,0} + (\rho^* V^m)_{,m} = 0$. We calculate $\Delta = |\det(g_{\mu\nu})|^{1/2}$ from (4.1), and find that

$\Delta = 1 + 2U + O_4$ (we cannot find a better approximation without specifying the g_{mn} more precisely). From (4.2) and the definition $\zeta = \rho c^2(1 + \Pi)$, og

$$\begin{aligned} \rho^* &= \rho(1 + 3U + \Upsilon + O_4), \\ \zeta &= \rho^* c^2(1 + \Pi - 3U - \Upsilon + O_4). \end{aligned} \tag{4.8}$$

Substituting this ρ^* into V(4.13) gives the postnewtonian continuity equation V(5.5). Substituting for ζ in (4.7) and multiplying by $(1 - 2\Upsilon - p/\rho^* c^2)(1 - \Pi + 3U + \Upsilon)$, og

$$\begin{aligned} &\rho^* D_t V^m + \rho^* D_t(4UV^m + cg_m) - \rho^* c V^m U_{,0} \\ &\quad + c^{-1} V^m p_{,0} + p_{,m}(1 - \Pi + 3U - \Upsilon - p/\rho^* c^2) \\ &\quad - \rho^* c^2(U_{,m} + 2UU_{,m} + 2\Upsilon U_{,m} + (1/2) g_{0,m} + V^r c^{-1} g_{r,m}) \\ &= \rho c^2 L^{-1} O_6. \end{aligned} \tag{4.9}$$

Eq.(4.9) is equivalent to Will's eq.(6.29) when the PPN parameters have their standard values ($\beta = \gamma = 1$, and the rest zero). To establish this, note that $g_m = -4\mathcal{V}_m + (1/2)(\mathcal{V}_m - \mathcal{W}_m)$, and that $(\mathcal{V}_m - \mathcal{W}_m)_{,n} = (\mathcal{V}_n - \mathcal{W}_n)_{,m} + L^{-1} O_5$ from (C11). The physical meaning of (4.7) and (4.9) is not obvious; in the next section we write them more perspicuously in terms of a momentum density.

Exercise 4.1 Give complete derivations (all the algebra!) of (4.7)–(4.9). ∎

The potential U is defined by V(2.4) in terms of the invariant density ρ. It is sometimes more convenient to use a potential U^* that is defined in terms of the conserved density ρ^*:

$$U^*(x) = kc^2 \int \rho^*(x^0, \boldsymbol{y}) |\boldsymbol{x} - \boldsymbol{y}|^{-1} d^3y. \tag{4.10}$$

As before, integrals are over all values of the integration variables, unless stated otherwise. From (4.8), VI(1.13), and VI(1.14), og

$$U^* = U + (1/2)\Phi_1 + 3\Phi_2 + O_6. \tag{4.11}$$

To this order, it is inconsequential whether we replace ρ or U by ρ^* or U^*, respectively, in Φ_1 and Φ_2. We shall usually drop the stars in such higher-order terms.

It follows from the continuity equation V(4.13) that $c(\rho^* V^m)_{,0} = -(\rho^* V^n)_{,n} V^m + c\rho^* V^m{}_{,0} = -(\rho^* V^n V^m)_{,n} + \rho^* V^n V^m{}_{,n} + c\rho^* V^m{}_{,0}$, and hence that

$$c(\rho^* V^m)_{,0} = -(\rho^* V^n V^m)_{,n} + \rho^* D_t V^m. \tag{4.12}$$

This allows us to write (4.9) in terms of the partial derivatives of $\boldsymbol{V}$. Also from V(4.13), we show that if f is any smooth function of the spacetime variables that falls off sufficiently rapidly at spatial infinity, then

$$(d/dt)\int f\rho^*(x^0,\boldsymbol{x})d^3x = \int \rho^* D_t f(x^0,\boldsymbol{x})d^3x. \tag{4.13}$$

When $f = V^m$, eq.(4.13) follows immediately from (4.12).

5. Conservation laws

A powerful and general way of investigating a system without solving its equations of motion is to look for conserved quantities. In this section we derive approximate conservation laws for the energy, momentum, and angular momentum of a gravitating ideal fluid.

We first recall, as usual, the special relativistic results. We assume that a system is associated with a tensor field Θ whose components $\Theta^{\mu\nu}$ in an inertial chart x satisfy ${\Theta^{\mu\nu}}_{,\nu} = 0$. At any instant $t = x^0/c$, we define

$$P^\mu(t) = c^{-1}\int \Theta^{\mu 0}(x)d^3x, \tag{5.1}$$

where $(x) = (x^0, \boldsymbol{x})$, and the integral is over all $\boldsymbol{x}$ and is assumed to converge. Provided that the $\Theta^{\mu m}$ fall off sufficiently rapidly for large values of $|\boldsymbol{x}|$, og $DP^\mu(t) = \int {\Theta^{\mu 0}}_{,0}(x)\,d^3x = -\int {\Theta^{\mu m}}_{,m}(x)\,d^3x = 0$ by the divergence theorem. It follows that the P^μ are constants, independent of t. They are the components in x of the total 4 momentum of the system, and we interpret cP^0 to be the energy. One can show – most easily by using the 4 dimensional version of the divergence theorem – that the P^μ transform like the components of a contravariant vector under Poincaré transformations.

In a similar manner, again assuming that the integral converges, we define

$$J^{\mu\nu}(t) = c^{-1}\int \left[\Theta^{\mu 0}(x)x^\nu - \Theta^{\nu 0}(x)x^\mu\right] d^3x. \tag{5.2}$$

Provided that Θ is symmetric ($\Theta^{\mu\nu} = \Theta^{\nu\mu}$), one proves that the $J^{\mu\nu}$ are constants, independent of t. They are the components in x of the total 4 angular momentum of the system.

The stress-momentum T of a gravitating, ideal fluid does not satisfy ${T^{\mu\nu}}_{,\nu} = 0$, but the more complicated, covariant equation (3.8) (which is equivalent to the vanishing of the covariant divergence ${T^{\mu\nu}}_{;\nu}$). One can however introduce new functions $\Theta^{\mu\nu} = T^{\mu\nu} + t^{\mu\nu}$, where the $t^{\mu\nu}$ are chosen so that ${\Theta^{\mu\nu}}_{,\nu}$ vanishes, to sufficient accuracy. One then defines conserved energy, momentum, and angular momentum in terms of the $\Theta^{\mu 0}$.

In what follows, we need the contravariant components $g^{\mu\nu}$ of the spacetime metric, which are the solutions of the equations $g^{\mu\nu}g_{\nu\pi} = \delta_{\mu\pi}$, where the $g_{\mu\nu}$ are given by (4.1). The easiest way to solve these equations is to guess the answer. You should verify that

$$\begin{gathered} g^{00} = -1 - 2U - 4U^2 - g_0 + O_6, \\ g^{m0} = g_m + O_5, \qquad g^{mn} = \delta_{mn}(1 - 2U) + O_4. \end{gathered} \tag{5.3}$$

Since $\Delta = 1 + 2U + O_4$ and $T^{\mu\nu} = g^{\mu\pi}T^{\nu}_{\pi}$, we have $T^{0\mu}{}_{,\mu} = (g^{0\nu}T^{\mu}_{\nu})_{,\mu} = (g^{00}T^{\mu}_{0})_{,\mu} + \rho c^2 L^{-1}O_5 = -(\Delta T^{\mu}_{0})_{,\mu} + \rho c^2 L^{-1}O_5$, and we show from (3.8) and (3.2) that

$$T^{0\nu}{}_{,\nu} = -U_{,0}T^{00} + \rho c^2 L^{-1}O_5, \qquad T^{m\nu}{}_{,\nu} = U_{,m}T^{00} + \rho c^2 L^{-1}O_4. \tag{5.4}$$

The terms in $U_{,\mu}$ on the right-hand side of (5.4) prevent us from defining conserved quantities in the easy manner of special relativity.

Exercise 5.1 Calculate the $\rho c^2 L^{-1}O_5$ terms in $T^{0\nu}{}_{,\nu}$ and the $\rho c^2 L^{-1}O_4$ terms in $T^{m\nu}{}_{,\nu}$ [cf.(5.9)]. ■

We recall that the potential U, eq.V(2.4), satisfies (C33):

$$U_{,mm} = -4\pi k\rho c^2 = -4\pi k T^{00} + L^{-2}O_4, \tag{5.5}$$

and it follows that $(4\pi k)^{-1}UU_{,mm} = \rho c^2 O_2$, $(4\pi k)^{-1}U_{,m}U_{,n} = \rho c^2 O_2$, and $(4\pi k)^{-1}U_{,m}U_{,0} = \rho c^2 O_3$.

We define functions $\Theta^{\mu\nu}$ by

$$\begin{aligned} \Theta^{00} &= T^{00} + (8\pi k)^{-1}U_{,r}U_{,r}, \\ \Theta^{m0} &= \Theta^{0m} = T^{m0} - (4\pi k)^{-1}U_{,m}U_{,0}, \\ \Theta^{mn} &= T^{mn} + (8\pi k)^{-1}(-U_{,r}U_{,r}\delta_{mn} + 2U_{,m}U_{,n}). \end{aligned} \tag{5.6}$$

Eqs.(5.4) and (5.5) imply that

$$\Theta^{0\nu}{}_{,\nu} = \rho c^2 L^{-1}O_5, \qquad \Theta^{m\nu}{}_{,\nu} = \rho c^2 L^{-1}O_4. \tag{5.7}$$

The total 4 momentum and total 4 angular momentum of the gravitating fluid are defined by (5.1) and (5.2) with $\Theta^{\mu\nu}$ given by (5.6). It follows from (5.7) that

$$DP^0 = dP^0/dt = \rho c^2 L^2 O_5, \qquad DP^m = dP^m/dt = \rho c^2 L^2 O_4, \tag{5.8}$$

and that P^0 and P^m are constant apart from terms of order $\rho c L^3 O_4$ and $\rho c L^3 O_3$, respectively. Similarly, using the symmetry of $\Theta^{\mu\nu}$, one shows that the $J^{\mu\nu}$ defined by (5.2) are approximately constant. We note that the Newtonian energy and momentum are of order $MV^2 = Mc^2O_2 = \rho c^2L^3O_2$ and $MV = McO_1 = \rho cL^3O_1$, respectively, and their time derivatives are of order $\rho c^3L^2O_3$ and $\rho c^2L^2O_2$. Eq.(5.8) therefore implies that the energy cP^0 and the momentum $\boldsymbol{P}$ are conserved in the Newtonian approximation – but not necessarily in the postnewtonian. We shall return to this point later.

We can write the momentum components in more explicit forms. From (3.2), (4.2), and (5.3), the components of the stress-momentum are (cf. Will, Table 4.1, and Soffel (4.7.9))

$$\begin{aligned} T^{00} &= \rho c^2(1+2U+\Pi+2\Upsilon)+\rho c^2O_4, \\ T^{m0} &= T^{0m} = \rho cV^m(1+2U+\Pi+2\Upsilon+p/\rho c^2)+\rho c^2O_5, \\ T^{mn} &= \rho V^mV^n(1+2U+\Pi+2\Upsilon+p/\rho c^2)+p(1-2U)\delta_{mn}+\rho c^2O_6. \end{aligned} \tag{5.9}$$

Using (5.5), og $(8\pi k)^{-1}U_{,r}U_{,r} = (8\pi k)^{-1}(UU_{,r})_{,r}+(1/2)\rho c^2U$, and the divergence theorem gives

$$(8\pi k)^{-1}\int U_{,r}U_{,r}(x)d^3x = (1/2)\int \rho c^2U(x)d^3x, \tag{5.10}$$

where again $(x) = (x^0, \boldsymbol{x})$.

Eqs.(5.1), (5.6), (5.9), and (5.10) imply that

$$P^0(t) = \int \rho c\big[1+5U/2+\Pi+2\Upsilon\big](x)\,d^3x+\rho cL^3O_4. \tag{5.11}$$

The P^m can be written in terms of the potentials $\mathcal{V}_r$ and $\mathcal{W}_r$ of Appendix C. We recall that the superpotential χ, defined by (C1), satisfies $\chi_{,0m} = \mathcal{V}_m - \mathcal{W}_m + O_5$, $\chi_{,mm} = -2U$, from (C11) and (C7). By repeated partial integration and use of (5.5), og

$$\begin{aligned} (4\pi k)^{-1}\int U_{,m}U_{,0}(x)\,d^3x &= -(4\pi k)^{-1}\int UU_{,0m}(x)d^3x \\ &= (8\pi k)^{-1}\int U(\mathcal{V}_m-\mathcal{W}_m)_{,rr}(x)d^3x+\rho c^2L^3O_5 \\ &= (8\pi k)^{-1}\int U_{,rr}(\mathcal{V}_m-\mathcal{W}_m)(x)d^3x+\rho c^2L^3O_5 \\ &= -(1/2)\int \rho c^2(\mathcal{V}_m-\mathcal{W}_m)(x)d^3x+\rho c^2L^3O_5. \end{aligned} \tag{5.12}$$

Substituting (5.9), (5.6), and (5.12) into (5.1) gives

$$P^m(t) = \int \Big[\rho V^m(1 + 2U + \Pi + 2\Upsilon + p/\rho c^2)$$

$$+ (1/2)\rho c(\mathcal{V}_m - \mathcal{W}_m)\Big](x)\, d^3x + \rho c L^3 O_5. \tag{5.13}$$

In terms of the conserved mass density ρ^*, eq.(4.8), og

$$P^0(t) = \int \rho^* c\big[1 - U/2 + \Pi + \Upsilon\big](x)\, d^3x + \rho c L^3 O_4, \tag{5.14}$$

$$P^m(t) = \int \big[\rho^* V^m(1 - U + \Pi + \Upsilon + p/\rho c^2) + (1/2)\rho c(\mathcal{V}_m - \mathcal{W}_m)\big](x)\, d^3x + \rho c L^3 O_5$$

$$= \int \big[\rho^* V^m(1 - U/2 + \Pi + \Upsilon + p/\rho c^2) - (1/2)\rho c \mathcal{W}_m\big](x)\, d^3x + \rho c L^3 O_5,$$

$$= \int \big[\rho^* V^m(1 + 3U + \Pi + \Upsilon + p/\rho c^2) + \rho c g_m\big](x)\, d^3x + \rho c L^3 O_5, \tag{5.15}$$

where we have used the definition $g_m = -(1/2)(7\mathcal{V}_m + \mathcal{W}_m)$, and the equation

$$\int \rho V^m U(x) d^3x = \int \rho c \mathcal{V}_m(x) d^3x, \tag{5.16}$$

which follows from the definition of $\mathcal{V}_m$, eq.(C12). As usual, we have written ρ for ρ^* when it makes no difference. Eqs.(5.14) and (5.15) agree with Will (4.108) and Soffel (4.7.22) (with $\beta = \gamma = 1$).

We have kept terms of order $\rho c L^3 O_3$ in the expressions for P^m even though we have not proved that these terms are conserved (eq.(5.8) states that $DP^m = \rho c^2 L^2 O_4$ – which implies only that P^m is constant except for terms of order $\rho c L^3 O_3$). By a happy chance, it turns out that the terms of order $\rho c L^3 O_3$ are in fact conserved: we are going to prove that $DP^m = \rho c^2 L^2 O_6$. We emphasize that there is nothing inevitable about this: we guessed the expressions (5.6) (which determine the P^m) in order to satisfy the approximate equations (5.7).

One begins the proof by differentiating (5.15) with respect to t and using (4.13):

$$DP^m(t) = \int \rho^* D_t\big[V^m(1 + 3U + \Pi + \Upsilon + p/\rho c^2) + c g_m\big](x)\, d^3x + \rho c^2 L^2 O_6, \tag{5.17}$$

where D_t is again the material or comoving derivative with respect to the time: $D_t f = c f_{,0} + V^m f_{,m}$. We could now substitute for $\rho^* D_t V^m$ from the Euler equations, but it is easier and more elegant to first rewrite them in a form suggested by (5.17).

In (4.7) we substitute the expression (4.8) for ζ – except in the last term where we put $\zeta = \rho c^2(1+\Pi)$. We then multiply the equation by $c^{-2}(1+2U)$, and got

$$\rho^*(D_tV^m)(1+3U+\Pi+\Upsilon+p/\rho c^2)+\rho c D_t g_m$$
$$+4\rho V^m D_tU-\rho V^m cU_{,0}+c^{-1}V^m p_{,0}+p_{,m}(1+2U)-pU_{,m}$$
$$-\rho c^2(1+4U+\Pi+4\Upsilon)U_{,m}-\rho c^2(g_{0,m}/2+V^r c^{-1}g_{r,m})$$
$$=\rho c^2 L^{-1}O_6. \tag{5.18}$$

We add and subtract terms $\rho^* V^m D_t(1+3U+\Pi+\Upsilon+p/\rho c^2)$, eliminate $D_t\Upsilon$ by using (4.5), note that (3.15) implies that $c^2D_t\Pi = -pD_t(\rho^{-1})$, and find

$$\rho^* D_t\Big[V^m(1+3U+\Pi+\Upsilon+p/\rho c^2)+cg_m\Big]+p_{,m}$$
$$+2(pU)_{,m}-3pU_{,m}-\rho c^2(1+4U+\Pi+4\Upsilon)U_{,m}$$
$$-\rho c^2(g_{0,m}/2+V^r c^{-1}g_{r,m})=\rho c^2L^{-1}O_6. \tag{5.19}$$

Substituting for g_0, defined as in (4.1), og

$$\rho^* D_t\Big[V^m(1+3U+\Pi+\Upsilon+p/\rho c^2)+cg_m\Big]+p_{,m}+2(pU)_{,m}-\rho c^2U_{,m}$$
$$-3(pU_{,m}+\rho c^2\Phi_{4,m})-\rho c^2(4\Upsilon U_{,m}+2\Phi_{1,m})-2\rho c^2(UU_{,m}+\Phi_{2,m})$$
$$-\rho c^2(\Pi U_{,m}+\Phi_{3,m})-\rho c V^r g_{r,m}=\rho c^2L^{-1}O_6, \tag{5.20}$$

where the Φ_α are given by (C17)–(C20), and $g_m = -(1/2)(7\mathcal{V}_m + \mathcal{W}_m)$.

Eqs.(5.19) or (5.20) are equivalent to the Euler equations (4.7). They are still fairly complicated. However, if one integrates (5.20) over all $\boldsymbol{x}$, it simplifies remarkably. The partial derivative terms vanish, and we have $\int d^3x\rho c^2 U_{,m}(x) = -kc\int d^3x\int d^3y\rho(x)\rho(y)(x^m-y^m)|\boldsymbol{x}-\boldsymbol{y}|^{-3}=0$ (exchange $\boldsymbol{x}$ and $\boldsymbol{y}$, and note that the integral is invariant and also changes sign). In the same way, one shows that the other terms vanish, and og

$$\int \rho^* D_t\Big[V^m(1+3U+\Pi+\Upsilon+p/\rho c^2)+cg_m\Big](x)d^3x=\rho c^2L^2O_6, \tag{5.21}$$

which from (5.17) and (4.13) implies that

$$DP^m(t)=\rho c^2L^2O_6. \tag{5.22}$$

We can simplify some calculations by defining functions T^μ:

$$T^0=\rho^* c\Big[1-U/2+\Pi+\Upsilon\Big],$$
$$T^m=\rho^* V^m(1-U+\Pi+\Upsilon+p/\rho c^2)+(1/2)\rho c(\mathcal{V}_m-\mathcal{W}_m). \tag{5.23}$$

They are the integrands of (5.14) and the first integral of (5.15). One may regard cT^0 as the energy density, and T^m as the m component of the momentum density. We note that they vanish at points where ρ^* vanishes.

There is a relationship between the T^μ that will be important in our later discussion of centres of mass. From the continuity equation V(4.13) og

$$T^0{}_{,0} = -\big[\rho^* V^m(1 - U/2 + \Pi + \Upsilon)\big]_{,m} + \rho D_t(-U/2 + \Pi + \Upsilon). \tag{5.24}$$

We substitute for $D_t\Pi$ from (3.15) and for $D_t\Upsilon$ from (4.5) and show that

$$T^0{}_{,0} = -T^m{}_{,m} + (1/2)c(\rho_{,0}U - \rho U_{,0}) + (1/2)\big[\rho c(\mathcal{V}_m - \mathcal{W}_m)\big]_{,m} + \rho c L^{-1} O_5. \tag{5.25}$$

Exercise 5.2 Derive (5.25). ∎

The T^μ are closely related to the $\Theta^{\mu 0}$ defined by (5.6) and (5.9). From (4.8) and (5.5), og

$$\begin{aligned} \Theta^{00} &= cT^0 + (8\pi k)^{-1}(UU_{,r})_{,r} + \rho c^2 O_4, \\ \Theta^{m0} &= cT^m - (1/2)\rho c^2(\mathcal{V}_m - \mathcal{W}_m) - (4\pi k)^{-1}U_{,m}U_{,0} + \rho c^2 O_5. \end{aligned} \tag{5.26}$$

From (5.1) and (5.12), we check that $P^m(t) = \int T^m(x)\, d^3x + \rho c L^3 O_5$.

Exercise 5.3 Use (5.25) and (5.26) to calculate $\Theta^{\mu 0}{}_{,\mu}$. [The result should agree with the first of eqs.(5.7).] ∎

Exercise 5.4 Write the components $J^{\mu\nu}$ of the angular momentum, eqs.(5.2), in terms of the T^μ, and calculate DJ^{m0}. ∎

The Euler equations (5.20) can be written in terms of T^m. We note that for any function f of x one has $\rho^* D_t f = \rho^* c f_{,0} + \rho^* V^r f_{,r} = (\rho^* c f)_{,0} - c\rho^*_{,0} f + \rho^* V^r f_{,r} = (\rho^* c f)_{,0} + (\rho^* V^r f)_{,r}$. Taking $f = V^m(1 + 3U + \Pi + \Upsilon + p/\rho c^2) + c g_m$, it follows that the first term of (5.20) is equal to $c\pi^m{}_{,0} + (V^r \pi^m)_{,r}$, where

$$\pi^m = \rho^* V^m(1 + 3U + \Pi + \Upsilon + p/\rho c^2) + \rho c g_m = T^m + 4\rho(V^m U - c\mathcal{V}_m). \tag{5.27}$$

This gives the "quasi-Newtonian" form of the Euler equations, which was first derived in [Chandrasekhar 69] (see also [Caporali 81], and Soffel (4.7.14) and Appendix A.4 – eq.(5.27) is equivalent to Soffel's (4.7.15) with $\gamma = 1$). By writing π^m in terms of Θ^{m0} from (5.26), one shows that the quasi-Newtonian Euler equations are equivalent to an improved version of the second of eqs.(5.7).

An important difference between the T^μ (or the π^m) and the $\Theta^{\mu 0}$ is that the former vanish where the density vanishes while the latter do not. This suggests (but does not prove!) that the T^μ should be used in the definition of the momentum of material bodies (cf. VIII, Section 4).

Exercise 5.5 Calculate $\Theta^{m\mu}{}_{,\mu}$ in the manner just outlined, and so derive the improved form of (5.7). ■

6. Postnewtonian field equations

The field equations that we derived in Chapter VI, Section 2 can be expressed more elegantly in terms of the stress-momentum of the sources. Instead of the contravariant stress-momentum (3.2), we need the covariant stress-momentum, whose components are $T_{\mu\nu} = pg_{\mu\nu} + (\zeta + p)v_\mu v_\nu c^{-2}$. From (4.2) og $v_0 c^{-1} = -1 + U - \Upsilon + O_4, v_m c^{-1} = V^m c^{-1} + O_3$, and the postnewtonian $g_{\mu\nu}$ are given by (4.1). It follows that (cf. (5.9))

$$\begin{gathered} T_{00} = \rho c^2(1 - 2U + \Pi + 2\Upsilon) + \rho c^2 O_4, \\ T_{m0} = -\rho c V^m + \rho c^2 O_3, \qquad T_{mn} = \rho V_m V_n + p\delta_{mn} + \rho c^2 O_4. \end{gathered} \tag{6.1}$$

Since $g^{\mu\nu}v_\mu v_\nu = v_\mu v^\mu = -c^2$, og $T_{\mu\nu}g^{\mu\nu} = 3p - \zeta$. The field equations VI(2.17) can be written as

$$\begin{aligned} \partial_r\partial_r g_{00} &= -4\partial_r U\partial_r U - 16\pi k(T_{00} - (1/2)g_{00}T_{\mu\nu}g^{\mu\nu} + \rho c^2 U) + L^{-2}O_6, \\ \partial_r\partial_r g_{m0} &= -\partial_0\partial_m U - 16\pi k T_{m0} + L^{-2}O_5, \\ \partial_r\partial_r g_{mn} &= 8\pi k g_{mn}T_{\mu\nu}g^{\mu\nu} + L^{-2}O_4, \end{aligned} \tag{6.2}$$

where we have used $g_{00} = -1 + 2U + O_4, g_{mn} = \delta_{mn}(1 + 2U) + O_4$.

7. Postinertial charts and field equations

Postinertial charts, which were introduced in the last chapter, have two important properties. The postnewtonian metric has the same functional form VI(3.9) in all of them, and any two of them are related by a semipoincaré transformation. We recall that a semipoincaré transformation is an approximate Poincaré transformation which is valid if the relative speed of the charts that it relates is small compared to the speed of light. More precisely, a semiboost is defined by VI(1.1), and a semipoincaré transformation is the composition of a semiboost with any combination of spatial rotations, shifts of origin, space reflection, and time reversal.

Any postinertial chart is related to a postnewtonian chart by a transformation of the form VI(3.1). Two postnewtonian charts are related by the composition of a semipoincaré transformation and a transformation VI(2.10) of the time coordinate. One advantage of postinertial charts over postnewtonian is that the additional transformation of the time coordinate is unnecessary.

The results derived in Sections 4 and 5 for postnewtonian charts are valid, with only trivial changes, for postinertial charts. We have already noted that if the chart x is postinertial, then in the expressions (4.1) for the postnewtonian metric one must replace g_μ by $g^\#_\mu$, defined by VI(3.11). The same substitution must be made in the other equations of Section 4, and in eqs.(5.1)–(5.14) (when g_μ does not appear in an equation, it is correct as it stands).

Eqs.(5.15)–(5.17) do not require any change (g_μ is not replaced by $g^\#_\mu$). In (5.18), which is derived from (4.7), one must replace g_μ by $g^\#_\mu$. However, one can then write $g^\#_0 = g_0 + 2\partial_0\theta + O_6$ and $g^\#_m = g_m + \partial_m\theta + O_5$, and one finds that the terms involving θ cancel: (5.18) remains valid with g_μ not replaced by $g^\#_\mu$. Similarly, (5.19)–(5.27) require no correction.

In summary, the postinertial versions of the postnewtonian eqs.(4.1)–(5.14) are found by replacing g_μ by $g^\#_\mu$; the postinertial versions of (5.15)–(5.27) are the same as the postnewtonian.

Postinertial field equations

In Section VI.3, we introduced an explicitly covariant form of the postnewtonian metric: eq.VI(3.16) gives the metric in a postinertial chart in terms of invariant functions α and ω and the components n_μ of a covector field. The invariant ω is not uniquely determined, and we concentrated on one of the simplest possibilities, $\omega = \omega_1 + \omega_2$ (see eqs.VI(3.19) and VI(3.15)). We derived field equations for $\alpha n_0 n_\mu$ and ω, and also for the field variables $\beta = \alpha n_0^2$ and $q_m = -n_m/n_0$. We are going to express these equations in terms of the stress-momentum of the sources, just as we did for the postnewtonian equations in Section 6. The results of this subsection will not be required until Chapter IX.

Since $g_{\mu\nu}v^\mu v^\nu = v_\mu v^\mu = -c^2$, og $g_{\mu\nu}T^{\mu\nu} = 3p - \zeta$, $v_\mu v_\nu T^{\mu\nu} = \zeta c^2$ from (3.2), and hence

$$\zeta = \rho c^2(1+\Pi) = c^{-2}v_\mu v_\nu T^{\mu\nu}, \qquad 3p = (g_{\mu\nu} + c^{-2}v_\mu v_\nu)T^{\mu\nu}. \tag{7.1}$$

One cannot express U in terms of the stress-momentum, but $U = -\omega + O_4$, from the requirement that VI(3.16) implies VI(3.10).

Using (5.9), we write the field equations VI(3.26) and VI(3.27) as

$$-\eta_{\mu\nu}\partial_\mu\partial_\nu(\beta q_m) = \eta_{\mu\nu}\partial_\mu\partial_\nu(\alpha n_0 n_m) = (\alpha n_0 n_m)_{,rr} + L^{-2}O_5 = 16\pi k T^{m0} + L^{-2}O_5. \tag{7.2}$$

Eqs.VI(3.31) and (5.9) give

$$\eta_{\mu\nu}\partial_\mu\partial_\nu\beta = -16\pi k\left[T^{00}(1+q_p q_p) + T^{pp} - 2T^{p0}q_p\right] - 2\beta\partial_m q_p \partial_m q_p + L^{-2}O_6. \quad (7.3)$$

Similarly, from VI(3.30) and (5.9),

$$\eta_{\mu\nu}\partial_\mu\partial_\nu\omega = 4\pi k\left[T^{00}(1+2q_p q_p) + T^{pp} - 4T^{p0}q_p\right] - 4\omega\partial_m q_p \partial_m q_p + L^{-2}O_6. \quad (7.4)$$

Any exact theory of gravity must reduce to the postnewtonian theory in the appropriate limit, and therefore be compatible with the approximate, postnewtonian (or postinertial) field equations. It is tempting to try to go in the opposite direction, and to guess exact field equations as generalizations of the postnewtonian ones. However, we should be cautious. All manner of generalizations are possible, and we do not know which to prefer. We shall discuss this further in Chapter IX.

VIII Motion of the Centre of Mass

1. Centre of mass in Newtonian theory

A *centre of mass* in Newtonian mechanics is a spatial point that is associated with an extended body. One can give a useful, partial description of the motion of the body by specifying how its centre of mass moves. One may define the orbit of a planet, for example, as the path followed by its centre of mass. In this section we outline the calculation of the motion of the centre of mass in Newtonian theory; the rest of the chapter deals with postnewtonian generalizations.

The total mass of a body B in Newtonian mechanics is given by $M_B^* = \int_B \rho^*(x)\, d^3x$, where $(x) = (x^0, \boldsymbol{x})$, ρ^* is the mass density, and the integral is over the volume of the body (which we denote more explicitly by $\Omega_B(t)$, where $t = x^0 c^{-1}$). Note that, as in our previous discussions of Newtonian fluid mechanics, we are using ρ^* and M^* instead of ρ and M. This will allow us to take over many equations unchanged into the postnewtonian theory, where ρ^* is the conserved mass density (and different from the invariant mass density ρ).

The coordinates $X_B^m(t)$ of the centre of mass of B at the instant $t = x^0 c^{-1}$ in a Galilean chart are defined by

$$M_B^* X_B^m(t) = \int_B \rho^*(x) x^m\, d^3x. \tag{1.1}$$

The velocity and acceleration of the centre of mass of B are defined to be $\boldsymbol{V}_B = D\boldsymbol{X}_B$ and $\boldsymbol{A}_B = D\boldsymbol{V}_B$, respectively. Since $c\rho^*_{,0} = -(\rho^* V^r)_{,r}$ from V(4.2), and $(\rho^* V^r)_{,r} x^m = (\rho^* V^r x^m)_{,r} - \rho^* V^m$, it follows from Gauss' theorem that

$$M_B^* V_B^m(t) = \int_B c\rho^*_{,0}(x) x^m\, d^3x = \int_B \rho^* V^m(x)\, d^3x. \tag{1.2}$$

Similarly, from the nonrelativistic Euler equation VII(2.2), og

$$M_B^* A_B^m(t) = c \int_B (\rho^* V^m)_{,0}(x)\, d^0x = \int_B w^m(x)\, d^3x, \tag{1.3}$$

where $\boldsymbol{w}$ is the force density acting on the fluid. The acceleration of the centre of mass is therefore equal to the total force on the body divided by its total mass; when the total force is zero, the centre of mass moves at constant velocity.

If the force density $\boldsymbol{w}$ is purely gravitational, og $w^m = \rho^* c^2 U^*_{,m}$, where U^* is defined by VII(4.10). We assume throughout this chapter that the matter is in the form of discrete bodies (as in the solar system, for example). We label the bodies with a subscript A, and write

$$\int_B w^m(x)\, d^3x = \sum_A \int_B \rho^* c^2 U^*_{A,m}(x)\, d^3x, \tag{1.4}$$

where the sum is over all A, and

$$U^*_A(x) = kc^2 \int_A \rho^*(x^0, \boldsymbol{y}) |\boldsymbol{x} - \boldsymbol{y}|^{-1}\, d^3y. \tag{1.5}$$

The term with $A = B$ in (1.4) vanishes:

$$kc^4 \int_B d^3x \int_B d^3y \rho^*(x^0, \boldsymbol{x}) \rho^*(x^0, \boldsymbol{y}) (x^m - y^m) |\boldsymbol{x} - \boldsymbol{y}|^{-3} = 0. \tag{1.6}$$

The proof is like that for VII(5.21): the integrand changes sign when $\boldsymbol{x}$ and $\boldsymbol{y}$ are exchanged, but this transformation leaves the integral invariant.

In evaluating the terms in (1.4) with $A \neq B$, we assume that the bodies are small in comparison with the distances that separate them. The crudest approximation would be to replace $\boldsymbol{x} - \boldsymbol{y}$ by $\boldsymbol{X}_B - \boldsymbol{X}_A$ (i.e. by the displacement from the centre of mass of A to that of B). To be more precise, we recall that for vectors $\boldsymbol{r}$ and $\boldsymbol{R}$, with $r = |\boldsymbol{r}|$, $R = |\boldsymbol{R}|$, and $R \gg r$, one has

$$|\boldsymbol{R} - \boldsymbol{r}|^{-1} = R^{-1} \sum_n P_n(\cos\theta)(r/R)^n, \tag{1.7}$$

where the sum is over $n \in \{0, 1, 2, \ldots\}$, and θ is the angle between $\boldsymbol{r}$ and $\boldsymbol{R}$ (defined by $\boldsymbol{r} \cdot \boldsymbol{R} = rR\cos\theta$, $0 \leq \theta \leq \pi$). The P_n are the Legendre polynomials, of which the first three are

$$P_0(u) = 1, \qquad P_1(u) = u, \qquad P_2(u) = (3/2)u^2 - 1/2. \tag{1.8}$$

The series (1.7) converges rapidly when $R \gg r$ (more precise statements can be found in the mathematics texts, or see Exercise 1.1 below). Substituting (1.8) with

$u = \cos\theta = \boldsymbol{r}\cdot\boldsymbol{R}/rR$ into (1.7), og

$$|\boldsymbol{R}-\boldsymbol{r}\,|^{-1} = R^{-1}\Big[1+\boldsymbol{r}\cdot\boldsymbol{R}R^{-2}+(3/2)(\boldsymbol{r}\cdot\boldsymbol{R})^2R^{-4}-(1/2)r^2R^{-2}+O(r^3R^{-3})\Big] \quad (1.9)$$

as $rR^{-1}\to 0$. One may call R^{-1} the *monopole* term, $\boldsymbol{r}\cdot\boldsymbol{R}R^{-3}$ the *dipole* term, and $(3/2)(\boldsymbol{r}\cdot\boldsymbol{R})^2R^{-5}-(1/2)r^2R^{-3}$ the *quadrupole* term – they derive from P_0, P_1, and P_2, respectively. The same names are used for the corresponding terms in the expansion of the gravitational potential as a series of Legendre polynomials.

Exercise 1.1 Show that

$$|\boldsymbol{R}-\boldsymbol{r}|^{-1} = R^{-1}\Big[1+(r/R)^2-2(r/R)\cos\theta\Big]^{-1/2}. \quad (1.10)$$

Use the binomial expansion to derive (1.7) from (1.10). What is the condition for the convergence of this expansion? ■

If $\boldsymbol{X}_A$ are the spatial coordinates of the centre of mass of body A, we define $\boldsymbol{z}=\boldsymbol{x}-\boldsymbol{X}_A$, and show from (1.1) that

$$\int_A \rho^*(x^0,\boldsymbol{X}_A+\boldsymbol{z})z^m\,d^3z = 0. \quad (1.11)$$

In (1.11), we have used $\boldsymbol{z}$ as integration variables, but have still written the domain of integration as A. Less ambiguously, we could have written it as $\Omega_A(t)$, where $\Omega_A(t)$ is the set of spatial points in body A at the instant t. The set of spatial points is invariant under change of integration variables.

To evaluate U^*_A, eq.(1.5), we write $\boldsymbol{y}=\boldsymbol{X}_A+\boldsymbol{z}$, $\boldsymbol{R}=\boldsymbol{x}-\boldsymbol{X}_A$, so that $|\boldsymbol{x}-\boldsymbol{y}| = |\boldsymbol{x}-\boldsymbol{X}_A-\boldsymbol{z}| = |\boldsymbol{R}-\boldsymbol{z}|$. Substituting from (1.9) and dropping the $O(r^3R^{-3})$ terms, og

$$U^*_A(x) = kc^2R^{-1}\int_A \rho^*(x^0,\boldsymbol{X}_A+\boldsymbol{z}) \Big[1+\boldsymbol{z}\cdot\boldsymbol{R}R^{-2}+(3/2)(\boldsymbol{z}\cdot\boldsymbol{R})^2R^{-4}-(1/2)\boldsymbol{z}\cdot\boldsymbol{z}R^{-2}\Big]\,d^3z. \quad (1.12)$$

Eq.(1.11) now gives

$$U^*_A(x) = kc^2M^*_AR^{-1}+(1/2)kc^2R^{-5}I^{pq}_A(t)(3\boldsymbol{R}^p\boldsymbol{R}^q-R^2\delta^{pq}), \quad (1.13)$$

where $\boldsymbol{R}^m$ is the m component of $\boldsymbol{R}$, and $R^m = |\boldsymbol{R}|^m$, $\delta^{pq}=\delta_{pq}$, and

$$I^{pq}_A(t) = \int_A \rho^*(x^0,\boldsymbol{X}_A+\boldsymbol{z})z^pz^q\,d^3z. \quad (1.14)$$

We note that $I_A^{pq}(t)$ is independent of $\boldsymbol{R}$. Formally, it is a function of $x^0 = ct$ and $\boldsymbol{X}_A$; but $\boldsymbol{X}_A$ is a function of t – the solution of $D^2\boldsymbol{X}_A = \boldsymbol{A}_A$, where $\boldsymbol{A}_A$ satisfies an equation like (1.3).

We define $\boldsymbol{w} = \boldsymbol{x} - \boldsymbol{X}_B$, $\boldsymbol{R}_{AB} = \boldsymbol{X}_B - \boldsymbol{X}_A$, so that $\boldsymbol{R} = \boldsymbol{x} - \boldsymbol{X}_A = \boldsymbol{w} + \boldsymbol{R}_{AB}$. We write $U_A^*(x) = F(\boldsymbol{R}, t) = F(\boldsymbol{R}_{AB} + \boldsymbol{w}, t)$, where $F(\boldsymbol{R}, t)$ is the right-hand side of (1.13), and expand as a power series in $\boldsymbol{w}$:

$$U_A^*(x) = F(\boldsymbol{R}_{AB}, t) + w^r F_{,r}(\boldsymbol{R}_{AB}, t) + (1/2) w^r w^s F_{,rs}(\boldsymbol{R}_{AB}, t) + (1/6) w^r w^s w^t F_{,rst}(\boldsymbol{R}_{AB}, t) + \dots \quad (1.15)$$

where $F_{,r} = (\partial/\partial \boldsymbol{R}^r_{AB})F$, etc. To lowest order, og

$$U_{A,m}^*(x) = F_{,m}(\boldsymbol{R}_{AB}, t) + w^s F_{,ms}(\boldsymbol{R}_{AB}, t) + (1/2) w^s w^t F_{,mst}(\boldsymbol{R}_{AB}, t) + \dots \quad (1.16)$$

In the last term it is sufficient to replace $F(\boldsymbol{R}_{AB}, t)$ by $kc^2 M_A^* R_{AB}^{-1}$, so that $F_{,mst}(\boldsymbol{R}_{AB}, t) = kc^2 M_A^* (\partial/\partial \boldsymbol{R}^m_{AB}) \left[-\delta_{st} R_{AB}^{-3} + 3\boldsymbol{R}^s_{AB} \boldsymbol{R}^t_{AB} R_{AB}^{-5}\right]$. From (1.13), (1.4), (1.3), (1.6), and (1.11), we then find that

$$\begin{aligned} M_B^* A_B^m(t) &= M_B^* (A_B^m)_{\text{Newt}}(t) \\ &= kc^4 \sum_{A,B} (\partial/\partial \boldsymbol{R}^m_{AB}) \big\{ M_A^* M_B^* R_{AB}^{-1} \\ &\quad + (1/2)\left[M_A^* Q_B^{pq}(t) + M_B^* Q_A^{pq}(t)\right] \boldsymbol{R}^p_{AB} \boldsymbol{R}^q_{AB} R_{AB}^{-5} \big\}, \end{aligned} \quad (1.17)$$

where

$$\begin{aligned} Q_A^{pq} &= 3I_A^{pq} - I_A^{rr}\delta^{pq} \\ &= \int_A \rho^*(x^0, \boldsymbol{X}_A + \boldsymbol{z})(3z^p z^q - z^r z^r \delta^{pq})\, d^3 z \end{aligned} \quad (1.18)$$

is the (p, q) component of the *quadrupole moment* of body A. We have written $\boldsymbol{A}_B = (\boldsymbol{A}_B)_{\text{Newt}}$ to emphasize that this is the acceleration calculated from Newtonian theory. Eq.(1.17) implies that the force on B due to A is the negative of the force on A due to B. It reduces to Will's eq.(6.18) if $Q_B^{pq} = 0$. We note that (1.17) is valid only if terms of order $(r/R)^3$ are negligible.

To solve (1.17) and find the paths of the centres of mass, one must usually resort to numerical integration. The problem is straightforward if the quadrupole moments are constant, or if they are determined at any instant by the positions of the bodies at that instant (the quasistatic approximation). In general, however, the quadrupole moments depend in a complicated way on the motions of the bodies.

Consider, for example, a pair of close binary stars that orbit each other in an eccentric orbit. The varying tidal forces cause internal oscillations of the stars, and one can expect that resonances between the oscillation periods and the orbital period will produce strange effects.

Newtonian energy of a set of discrete bodies

We end the section by deriving an expression for the energy of a set of discrete bodies of the kind that we have been discussing. From VII(5.14) and V(2.4), the energy is

$$E = cP^0(t) = \sum_B \int_B d^3x\, \rho^* c^2 (1 + \Pi + \Upsilon)(x)$$

$$- (1/2)kc^4 \sum_{A,B} \int_B d^3x \int_A d^3y\, \rho^*(x)\rho^*(y)|\boldsymbol{x} - \boldsymbol{y}|^{-1} + \rho c^2 L^3 O_4, \quad (1.19)$$

where ρ^* is now the conserved mass density, $(x) = (x^0, \boldsymbol{x})$, $(y) = (y^0, \boldsymbol{y})$. In all but the first term, one can of course replace ρ^* by ρ. We write $\boldsymbol{V} = \boldsymbol{V}_B + \boldsymbol{v}_B$, so that $\int_B d^3x \rho^* c^2 \Upsilon = (1/2)\int_B d^3x \rho^* |\boldsymbol{V}_B + \boldsymbol{v}_B|^2 = (1/2)\int_B d^3x \rho^* |\boldsymbol{V}_B|^2 + (1/2)\int_B d^3x \rho^* |\boldsymbol{v}_B|^2$, by (1.2). In the second term of (1.19), we use the monopole approximation when $A \neq B$, and find that

$$E = \sum_B \Big[M_B^* c^2 + (1/2) M_B^* |\boldsymbol{V}_B|^2$$

$$+ (1/2) \int_B d^3x\, \rho^* |\boldsymbol{v}_B|^2(x) + \int_B d^3x \rho^* c^2 \Pi(x)$$

$$- (1/2)kc^4 \int_B d^3x \int_B d^3y\, \rho^*(x)\rho^*(y)|\boldsymbol{x} - \boldsymbol{y}|^{-1} \Big]$$

$$- kc^4 \sum_{A<B} M_A^* M_B^* R_{AB}^{-1} + \rho c^2 L^3 O_4. \quad (1.20)$$

The last summation is over all A and B such that $A <$ B. The terms in (1.20) can be identified as rest energy, centre of mass kinetic energy, internal kinetic energy, internal energy, internal gravitational potential energy, and Newtonian gravitational interaction energy of the bodies A and B. Note that the conserved mass M_B^* (which is of course constant in time) plays a role similar to that of the empirical mass in gravitostatics.

2. Centre of mass in postnewtonian theory

There are several plausible ways of defining the centre of mass in the postnewtonian theory. The most obvious is to use the same equation as in the Newtonian theory.

We assume as usual that the sources consist of ideal fluid, but now interpret ρ^* in (1.1) to be the conserved mass density, and M_B^* to be the conserved mass of the body. We then find the velocity and acceleration of the centre of mass in the postnewtonian approximation. In this way, we can calculate many interesting postnewtonian effects in a straightforward manner. (The calculations are somewhat laborious, but the pain is no worse than from running ten kilometres or planting a few hundred trees.)

The main disadvantage of defining the centre of mass in terms of the conserved mass density is that the centre of mass of a body may accelerate even when there is no external force. The "self-acceleration" terms are usually very small, and in this section and the next we shall usually ignore them. In Section 4 we show that they can be eliminated by a more complicated definition of the centre of mass.

Since ρ^* satisfies V(4.13), which has exactly the same form as V(4.2), the expression (1.2) for the velocity of the centre of mass of a body is still valid. The acceleration of the centre of mass is found by differentiating (1.2). We use VII(4.13) and the divergence theorem, assume that ρ^* vanishes on the surface of the domain of integration B (more precisely $\Omega_B(t)$ – the set of spatial points in body B at time t), and find that

$$M_B^* A_B^m(t) = c\int_B (\rho^* V^m)_{,0}(x)\, d^3x = \int_B \rho^* D_t V^m(x)\, d^3x, \tag{2.1}$$

where D_t is the comoving derivative with respect to the time t, and $(x) = (x^0, \boldsymbol{x})$.

We write $D_t(V^m + 4UV^m) = (1+4U)D_tV^m + 4V^m D_tU$ in VII(4.9), and solve for D_tV^m by multiplying the equation by $1-4U$. We eliminate D_tV^m from (2.1), substitute $g_0 = 4\Phi_1 + 4\Phi_2 + 2\Phi_3 + 6\Phi_4 - 2U^2$ from VII(4.1) and $U^* = U + (1/2)\Phi_1 + 3\Phi_2 + O_6$ from VII(4.11), and get

$$\begin{aligned} M_B^* A_B^m(t) = \int_B \Big\{ & -\rho c^2 g_{m,0} - 3\rho c V^m U_{,0} - c^{-1} V^m p_{,0} - 4\rho V^m V^r U_{,r} \\ & - p_{,m}(1 - \Pi - U - \Upsilon - p/\rho c^2) \\ & + \rho^* c^2 \big[U^*_{,m} + (3/2)\Phi_{1,m} - \Phi_{2,m} + \Phi_{3,m} + 3\Phi_{4,m} - 4UU_{,m} \\ & + 2\Upsilon U_{,m} + V^r c^{-1}(g_{r,m} - g_{m,r})\big] + \rho c^2 L^{-1} O_6 \Big\}(x)\, d^3x. \end{aligned} \tag{2.2}$$

Only the $\rho^* c^2 U^*_{,m}$ and $p_{,m}$ terms are of Newtonian order. The rest are postnewtonian; one can transform them by means of the nonrelativistic Euler equations VII(4.4), or replace ρ^* by ρ in them, without significant error. (The difference between ρ and ρ^* in the postnewtonian theory is given by VII(4.8), and is of order ρO_2; it is therefore negligible in the Newtonian approximation.)

For convenience, we collect together the nonrelativistic equations VII(2.4), VII(2.5), VII(4.12), VII(3.15), VII(5.5), and (1.3)–(1.5) that will be used repeatedly, and often without comment, in the remainder of this chapter:

$$c(\rho)_{,0} = -(\rho V^n)_{,n}, \qquad D_t\rho = c(\rho)_{,0} + V^n\rho_{,n} = -\rho V^n{}_{,n},$$

$$\rho D_t V^m = \rho A^m = c(\rho V^m)_{,0} + (\rho V^n V^m)_{,n} = -p_{,m} + \rho c^2 U_{,m},$$

$$c^2 D_t \Pi = -p D_t(\rho^{-1}) = cp\rho^{-2}\rho_{,0} = -p\rho^{-1}V^n{}_{,n},$$

$$U_{,mm} = -4\pi k\rho c^2,$$

$$M_B A_B^m = \{\rho A_B^m\}_B = \sum_{A\neq B}\{\rho c^2 U_{A,m}\}_B,$$
$$\{\rho(A^m - A_B^m)\}_B = \{\rho c^2 U_{B,m}\}_B. \tag{2.3}$$

We have written (2.3) as Newtonian equations, without postnewtonian error terms, and have omitted the optional stars. We have introduced the notation $\{f\}_B = \int_B f(x)\,d^3x$, for any function f of the variables $(x^0, \boldsymbol{x})$. (The conserved mass of body A is $M_A^* = \{\rho^*\}_A$, for example.) Although $\{f\}_B$ is a function of $t = x^0c^{-1}$, we shall usually not show this explicitly.

The contribution of the "Newtonian" term $\rho^* c^2 U^*_{,m}$ in (2.2) can be evaluated as in the last section; it is again written as $M_B^*(A_B^m)_{\text{Newt}}$, and is given by (1.17). The other "Newtonian" term, $-p_{,m}$, vanishes on integration. We shall calculate the remaining, truly postnewtonian terms in the crude, monopole approximation – which is almost always adequate for solar-system motions. This means that we replace $\boldsymbol{x} - \boldsymbol{y}$ by $\boldsymbol{X}_B - \boldsymbol{X}_A = \boldsymbol{R}_{AB}$ when $A \neq B$, where $\boldsymbol{x}$ and $\boldsymbol{y}$ are the coordinates of spatial points in the bodies B and A, respectively. As before, $\boldsymbol{X}_B$ and $\boldsymbol{X}_A$ are the coordinates of the centres of mass of B and A, and $\boldsymbol{R}_{AB}$ is the displacement from the centre of mass of A to that of B. In what follows, we usually omit the error terms in the expressions for the postnewtonian force contributions. However, it is important to remember that they are there, and that they consist of the postnewtonian $\rho c^2 L^2 O_6$ together with the term associated with the monopole approximation (the neglected dipole and quadrupole contributions, etc.).

The calculation of the postnewtonian terms in (2.2) is tedious but not difficult. The results are summarized near the end of the section (*Calculation of acceleration terms*). We write the acceleration of the centre of mass of B in the form

$$\boldsymbol{A}_B = (\boldsymbol{A}_B)_{\text{Newt}} + (\boldsymbol{A}_B)_{\text{self}} + (\boldsymbol{A}_B)_{\text{oth}}. \tag{2.4}$$

The first term has already been discussed. The "self- acceleration" $(\boldsymbol{A}_B)_{\text{self}}$ is a sum of integrals over the body B only (i.e. over the spatial region $\Omega_B(t)$ at the instant t when the acceleration is being calculated). The remaining term $(\boldsymbol{A}_B)_{\text{oth}}$ is a sum of integrals over one or more *other* bodies – and perhaps also over B (i.e. the domain of integration may be $\Omega_A(t)$ or $\Omega_A(t)\times\Omega_B(t)$ or $\Omega_A(t)\times\Omega_B(t)\times\Omega_C(t)$, etc., where $A\neq B$). To make this unambiguous, one has to adopt a convention that all potentials have to be expressed as integrals over the sources. In addition, we have rather arbitrarily included the time derivative term $-\{c^{-1}V^m p_{,0}\}_B$ in $(\boldsymbol{A}_B)_{\text{self}}$ (a more satisfactory way of dealing with this term will be given in Section 4).

The small term $(\boldsymbol{A}_B)_{\text{oth}}$ is calculated in the monopole approximation. As before, we define $\boldsymbol{R}^m_{AB}=X^m_B-X^m_A$, $R_{AB}=|\boldsymbol{R}_{AB}|$, $R^m_{AB}=|\boldsymbol{R}_{AB}|^m$. All integrals are evaluated at the same instant $t=x^0c^{-1}$, and for brevity we write $(x)=(x^0,\boldsymbol{x})$, $(y)=(x^0,\boldsymbol{y})$, $(z)=(x^0,\boldsymbol{z})$.

For any body A, we define

$$I^{rs}_A(t)=\int_A\rho(x)(x^r-X^r_A)(x^s-X^s_A)\,d^3x, \tag{2.5}$$

$$\begin{aligned}\beta^{rs}_A(t)&=\int_A\rho(x)(V^r-V^r_A)(V^s-V^s_A)\,d^3x\\&=\{\rho(V^r-V^r_A)(V^s-V^s_A)\}_A,\end{aligned} \tag{2.6}$$

in the notation introduced after (2.3). Eq.(2.5) is equivalent to (1.14); in what follows we often drop the argument $t=x^0c^{-1}$.

We prove at the end of this section that

$$\begin{aligned}D^2I^{rs}_A&=2\beta^{rs}_A+2\delta^{rs}\{p\}_A\\&\quad-kc^4\int_A d^3x\int_A d^3y\,\rho(x)\rho(y)(x^r-y^r)(x^s-y^s)|\boldsymbol{x}-\boldsymbol{y}|^{-3}.\end{aligned} \tag{2.7}$$

Setting $r=s$, og

$$D^2I^{rr}_A=2\beta^{rr}_A+6\{p\}_A-kc^4\int_A d^3x\int_A d^3y\,\rho(x)\rho(y)|\boldsymbol{x}-\boldsymbol{y}|^{-1}. \tag{2.8}$$

We define U_A and $\mathcal{U}_{rsA}$ by restricting the domain of integration in (C2) and (C8) to the spatial region occupied by body A (cf. the definition of U^*_A in (1.5)). We can then write the last term in (2.7) as $-c^2\{\rho\mathcal{U}_{rsA}\}_A$, and the last term in (2.8) as $-c^2\{\rho U_A\}_A$.

Substituting the equations from the end of the section into (2.2) and using (2.7) and (2.8) og

$$\begin{aligned}M_B(A^m_B)_{\text{oth}}=\sum_{A\neq B}kc^2R^{-3}_{AB}\big[&M_AM_B(3\boldsymbol{V}_A\cdot\boldsymbol{R}_{AB}-4\boldsymbol{V}_B\cdot\boldsymbol{R}_{AB})(V^m_A-V^m_B)\\&+3(M_B\beta^{mr}_A+M_A\beta^{mr}_B)\boldsymbol{R}^r_{AB}+(1/2)M_AD^2I^{mr}_B\boldsymbol{R}^r_{AB}\big]\end{aligned}$$

$$+ \sum_{A \neq B} kc^2 R_{AB}^{-3} \boldsymbol{R}_{AB}^m \Big[M_A M_B (-2\boldsymbol{V}_A \cdot \boldsymbol{V}_A - \boldsymbol{V}_B \cdot \boldsymbol{V}_B + 4\boldsymbol{V}_A \cdot \boldsymbol{V}_B)$$

$$+ (3/2) M_A M_B R_{AB}^{-2} (\boldsymbol{R}_{AB} \cdot \boldsymbol{V}_A)^2 + 3M_A \{p\}_B + 6M_B \{p\}_A - M_B c^2 \{\rho\Pi\}_A$$

$$- M_B D^2 I_A^{rr} + (3/2) M_B R_{AB}^{-2} \boldsymbol{R}^r_{AB} \boldsymbol{R}^s_{AB} \beta_A^{rs} - (1/2) M_A D^2 I_B^{rr} \Big]$$

$$+ \sum_{A \neq B} \sum_{C \neq A,B} k^2 c^6 M_A M_B M_C R_{AB}^{-3} \boldsymbol{R}_{AB}^m \Big[R_{AC}^{-1} + 4R_{BC}^{-1} + (1/2) R_{AC}^{-3} \boldsymbol{R}_{AB} \cdot \boldsymbol{R}_{AC} \Big]$$

$$+ \sum_{A \neq B} \sum_{C \neq A,B} (7/2) k^2 c^6 M_A M_B M_C R_{AB}^{-1} R_{AC}^{-3} \boldsymbol{R}_{AC}^m$$

$$+ \sum_{A \neq B} k^2 c^6 R_{AB}^{-4} \boldsymbol{R}_{AB}^m M_A M_B (5M_B + 4M_A). \tag{2.9}$$

The first bracket contains velocity dependent terms – some proportional to the relative velocity of the centres of mass $\boldsymbol{V}_A - \boldsymbol{V}_B$, and others depending on the internal velocities through the β^{mn}. The second bracket is a correction to the Newtonian, inverse-square law. The terms involving summations over A, B, and C are three-body forces – the forces on body B due to A and C together, where A, B, and C are distinct (these forces vanish if only two bodies are present). The last summation contains terms involving $M_A^2 M_B$ or $M_A M_B^2$ (we can regard them as degenerate three-body forces).

The Nordtvedt effect

The expression (2.9) for the gravitational force simplifies greatly if the velocities vanish at some instant. This is an unlikely occurrence for most real systems, but by supposing it we can gain some insight. We assume that $\boldsymbol{V}_A$, β_A^{mn}, and $D^2 I_A^{mn}$ vanish, use (2.8), neglect the small 3 body and degenerate 3 body terms, and find that

$$M_B (A_B^m)_{\text{oth}} = \sum_{A \neq B} kc^2 R_{AB}^{-3} \boldsymbol{R}_{AB}^m \Big[M_B c^2 \{\rho U_A\}_A + (1/2) M_A c^2 \{\rho U_B\}_B - M_B c^2 \{\rho\Pi\}_A \Big]. \tag{2.10}$$

It follows that there is a term in $(\boldsymbol{A}_B)_{\text{oth}}$ which is proportional to $\{\rho U_B\}_B / M_B c^2$: the acceleration of body B depends on the ratio of its gravitational energy to its rest energy. This should not surprise us because the conserved mass M_B^* that we have used in calculating $(\boldsymbol{A}_B)_{\text{Newt}}$ (cf. (2.4) and (1.17)) is not the same as the passive gravitational mass $\mathcal{M}_B$ that we introduced earlier, eq.V(2.6). If we substitute for M_B^* in terms of $\mathcal{M}_B$ in the monopole part $-kc^2 M_A^* M_B^* R_{AB}^{-3} \boldsymbol{R}_{AB}^m$ of $M_B^* (\boldsymbol{A}_B)_{\text{Newt}}$, and use VII(4.8) and (2.8), we find that the term proportional to $\{\rho U_B\}_B / M_B c^2$ in (2.10) is cancelled.

It is, of course, unsatisfactory to patch up the theory in this way. We should be more systematic and introduce a new mass density whose integral is the passive gravitational mass. Also, since the Earth and Moon are not at rest, we must use the general equations of motion instead of static approximations such as (2.10). The acceleration of the new centre of mass (calculated with the new mass density) will then, with any luck, not contain the $\{\rho U_B\}_B$ term. This will be discussed further in Section 4.

The question of which mass density to use is of more than academic interest. The Earth-Moon distance can be measured so precisely that the gravitational self energy terms $\{\rho U_B\}_B$ are in principle measurable. However, the interpretation of the measured results is not trivial. First one must choose a definition of the mass density and calculate (using suitable planetary models) the orbits of the Earth and Moon – the paths followed by their centres of mass. Then one must carefully analyze how the centres of mass are related to the positions of the laser on the Earth and the reflectors on the Moon that are used to measure the Earth-Moon distance. A proper analysis is complex and difficult. The reader should consult [Barker 86], [Brumberg 89], the books by Will and Soffel, and the references therein.

The possible distortion of the Earth-Moon orbit due to the gravitational self-energy terms is called the *Nordtvedt effect*. It is measurable, but not *directly* measurable. The purpose of our discussion has been to outline the arguments necessary for its elucidation.

The Einstein-Infeld-Hoffmann equation

If one neglects the terms involving p, Π, β_A^{mn}, and I_A^{mn} in (2.9), and also the self-acceleration in (2.4), eq.(2.2) reduces to the EIH (or *Einstein-Infeld-Hoffmann*) equation:

$$
\begin{aligned}
m_B A_B^m = & \sum_{A\neq B} kc^2 R_{AB}^{-3}\left[m_A m_B(3\boldsymbol{V}_A\cdot\boldsymbol{R}_{AB} - 4\boldsymbol{V}_B\cdot\boldsymbol{R}_{AB})(V_A^m - V_B^m)\right] \\
& + \sum_{A\neq B} kc^2 R_{AB}^{-3}\boldsymbol{R}_{AB}^m\Big[m_A m_B(-1 - 2\boldsymbol{V}_A\cdot\boldsymbol{V}_A - \boldsymbol{V}_B\cdot\boldsymbol{V}_B + 4\boldsymbol{V}_A\cdot\boldsymbol{V}_B) \\
& \qquad + (3/2)m_A m_B R_{AB}^{-2}(\boldsymbol{R}_{AB}\cdot\boldsymbol{V}_A)^2\Big] \\
& + \sum_{A\neq B}\sum_{C\neq A,B} k^2c^6 m_A m_B m_C R_{AB}^{-3}\boldsymbol{R}_{AB}^m\left[R_{AC}^{-1} + 4R_{BC}^{-1} + (1/2)R_{AC}^{-3}\boldsymbol{R}_{AB}\cdot\boldsymbol{R}_{AC}\right] \\
& + \sum_{A\neq B}\sum_{C\neq A,B} (7/2)k^2c^6 m_A m_B m_C R_{AB}^{-1}R_{AC}^{-3}\boldsymbol{R}_{AC}^m \\
& + \sum_{A\neq B} k^2c^6 R_{AB}^{-4}\boldsymbol{R}_{AB}^m m_A m_B(5m_B + 4m_A).
\end{aligned}
\tag{2.11}
$$

This is the equation used for calculating the motions of planets, with large computers. The constant masses m_B are not quite the same as the M_B^* (see Section 4 for more details). In practice, they are determined empirically to give the best fit to the observational data. The Newtonian quadrupole term is omitted from (2.11); it can be added if required. Eq.(2.11) is equivalent to Soffel's (4.1.14) with $\beta = \gamma = 1$.

The EIH equations were originally derived from the Einstein theory of gravity (by EIH, and also by Eddington and Clark in 1938). The calculations are summarized in [Misner 73], Exercise 39.15. The equations are also used for dense bodies such as neutron stars, where they require further justification because the postnewtonian approximation is of doubtful validity (Will, Section 11.3).

It is sometimes convenient to derive the EIH equations from a Lagrangian: see Will, eq.(11.62), and Soffel, eqs.(4.1.13) and (4.7.51) (the latter includes the spin-dependent terms discussed in Section 3 of this chapter).

Exercise 2.1 Make a rough calculation of the magnitude of the potential U in a neutron star, and estimate the error in using the postnewtonian approximation to describe such a star. [Take the mass of the star to be about the solar mass, and the radius to be approximately 10km.] ■

Exercise 2.2 Use (2.11) to calculate the postnewtonian contribution to the acceleration of the Earth due to Jupiter. [Only order-of-magnitude estimates of the sizes of the individual terms are required, but you should describe how their directions depend on the velocities and relative positions of the bodies. The mass of Jupiter is 1.90×10^{27}kg, its orbit about the Sun is almost circular with a radius of 7.78×10^8km, and its orbital period is 11.9 years.] ■

Exercise 2.3 Assume that the postnewtonian metric VI(2.12) is produced by a set of spatially small bodies. Find $g_{\mu\nu}(x)$ at points outside these bodies, assuming that $\Phi_3 = \Phi_4 = 0$ and that $U, \mathcal{V}_m, \mathcal{W}_m, \Phi_1$, and Φ_2 can be calculated in the monopole approximation (cf. VI(1.12)–VI(1.16)). Use the geodesic equation, V(6.3), to find the acceleration of a small body B at x. Compare your result with that given by (2.11). [Neglect the gravitational field produced by B. Recall that $\boldsymbol{V}$ in V(6.3) is regarded as a function of t only.] ■

The acceleration term $(\boldsymbol{A}_B)_{\text{self}}$ in (2.4) can be derived, in the same way as $(\boldsymbol{A}_B)_{\text{oth}}$, from the equations near the end of this section. It is usually small.

Exercise 2.4 Derive the expression for $(\boldsymbol{A}_B)_{\text{self}}$. [This is a sadistic problem. Just make sure that you understand how to do it, but do not grind through all the algebra. A better way is given in Section 4]. ■

Calculation of acceleration terms

We outline the calculation of the terms in (2.2) that contribute to $(A_B)_{6th}$, defined as in (2.4). Since they are all of postnewtonian order, one can use the nonrelativistic equations (2.3) in simplifying them. It is also permissible to replace M^* by M and ρ^* by ρ, and to write U_A instead of U_A^* in (1.5), etc. A summation over A is to be taken over all values of A, unless stated otherwise: one has $U = \sum_A U_A$, for example. The derivations are straightforward, and we shall be terse; in some cases we shall simply quote the result. However, the reader should diligently work through everything: it is the only way to find out which equations are incorrect. Note that postnewtonian error terms are not shown in the following calculations.

We perform the first calculation very explicitly:

$$
\begin{aligned}
\{p_{,m}U\}_B &= \sum_A \int_B d^3x \int_A d^3y\, kc^2 p_{,m}(x)\rho(y)|\boldsymbol{x}-\boldsymbol{y}|^{-1} \\
&= \sum_A \int_B d^3x \int_A d^3y\, kc^2 p(x)\rho(y)(x^m - y^m)|\boldsymbol{x}-\boldsymbol{y}|^{-3} \\
&= \int_B d^3x \int_B d^3y\, kc^2 p(x)\rho(y)(x^m - y^m)|\boldsymbol{x}-\boldsymbol{y}|^{-3} \\
&\quad + \sum_{A\neq B} kc^2 R_{AB}^{-3}\boldsymbol{R}_{AB}^m M_A\{p\}_B. \qquad (2.12)
\end{aligned}
$$

In a similar manner, og

$$
\begin{aligned}
\{\rho V^m V^r U_{,r}\}_B &= -\int_B d^3x \int_B d^3y\, kc^2 \rho V^m V^r(x)\rho(y)(x^r - y^r)|\boldsymbol{x}-\boldsymbol{y}|^{-3} \\
&\quad - \sum_{A\neq B} kc^2 R_{AB}^{-3}\boldsymbol{R}_{AB}^r M_A\{\rho V^m V^r\}_B, \qquad (2.13)
\end{aligned}
$$

$$
\begin{aligned}
\{\rho V^r V^r U_{,m}\}_B &= 2\{\rho c^2 \Upsilon U_{,m}\}_B \\
&= -\int_B d^3x \int_B d^3y\, kc^2 \rho V^r V^r(x)\rho(y)(x^m - y^m)|\boldsymbol{x}-\boldsymbol{y}|^{-3} \\
&\quad - \sum_{A\neq B} kc^2 R_{AB}^{-3}\boldsymbol{R}_{AB}^m M_A\{\rho V^r V^r\}_B. \qquad (2.14)
\end{aligned}
$$

Using VI(1.13)–(1.16), og

$$
\begin{aligned}
\{\rho \Phi_{1,m}\}_B &= -\int_B d^3x \int_B d^3y\, k\rho(x)\rho V^r V^r(y)(x^m - y^m)|\boldsymbol{x}-\boldsymbol{y}|^{-3} \\
&\quad - \sum_{A\neq B} kR_{AB}^{-3}\boldsymbol{R}_{AB}^m M_B\{\rho V^r V^r\}_A, \qquad (2.15)
\end{aligned}
$$

$$\{\rho\Phi_{3,m}\}_B = -\int_B d^3x \int_B d^3y\, kc^2\rho(x)\rho\Pi(y)(x^m - y^m)|\boldsymbol{x}-\boldsymbol{y}|^{-3}$$

$$- \sum_{A\neq B} kc^2 R_{AB}^{-3} \boldsymbol{R}_{AB}^m M_B \{\rho\Pi\}_A, \tag{2.16}$$

$$\{\rho\Phi_{4,m}\}_B = -\int_B d^3x \int_B d^3y\, k\rho(x)p(y)(x^m - y^m)|\boldsymbol{x}-\boldsymbol{y}|^{-3}$$

$$- \sum_{A\neq B} kR_{AB}^{-3} \boldsymbol{R}_{AB}^m M_B \{p\}_A. \tag{2.17}$$

The Φ_2 term is more complicated because it involves the potential U:

$$\{\rho\Phi_{2,m}\}_B = -\sum_A \int_B d^3x \int_A d^3y\, kc^2\rho(x)\rho U(y)(x^m - y^m)|\boldsymbol{x}-\boldsymbol{y}|^{-3}$$

$$= -\int_B d^3x \int_B d^3y\, kc^2\rho(x)\rho U(y)(x^m - y^m)|\boldsymbol{x}-\boldsymbol{y}|^{-3}$$

$$- \sum_{A\neq B} kc^2 R_{AB}^{-3} \boldsymbol{R}_{AB}^m M_B \{\rho U\}_A.$$

The first term is

$$-\int_B d^3x \int_B d^3y \int_B d^3z\, k^2c^4\rho(x)\rho(y)\rho(z)|\boldsymbol{y}-\boldsymbol{z}|^{-1}(x^m - y^m)|\boldsymbol{x}-\boldsymbol{y}|^{-3}$$

because

$$-\sum_{C\neq B} M_C R_{CB}^{-1} \int_B d^3x \int_B d^3y\, k^2c^4\rho(x)\rho(y)(x^m - y^m)|\boldsymbol{x}-\boldsymbol{y}|^{-3} = 0.$$

We also have

$$\{\rho U\}_A = \int_A d^3y \int_A d^3z\, kc^2\rho(y)\rho(z)|\boldsymbol{y}-\boldsymbol{z}|^{-1} + \sum_{C\neq A} kc^2 M_A M_C R_{AC}^{-1},$$

and hence

$$\{\rho\Phi_{2,m}\}_B =$$

$$-\int_B d^3x \int_B d^3y \int_B d^3z\, k^2c^4\rho(x)\rho(y)\rho(z)|\boldsymbol{y}-\boldsymbol{z}|^{-1}(x^m - y^m)|\boldsymbol{x}-\boldsymbol{y}|^{-3}$$

$$-\sum_{A\neq B}\sum_{C\neq A,B} k^2c^4 M_A M_B M_C R_{AC}^{-1} R_{AB}^{-3} \boldsymbol{R}_{AB}^m - \sum_{A\neq B} k^2c^4 M_A M_B^2 R_{AB}^{-4} \boldsymbol{R}_{AB}^m$$

$$-\sum_{A\neq B} k^2c^4 R_{AB}^{-3} \boldsymbol{R}_{AB}^m M_B \int_A d^3y \int_A d^3z\, \rho(y)\rho(z)|\boldsymbol{y}-\boldsymbol{z}|^{-1}. \tag{2.18}$$

In a similar manner, one finds

$$\{\rho U U_{m}\}_{B} =$$

$$- \int_B d^3x \int_B d^3y \int_B d^3z\, k^2c^4 \rho(x)\rho(y)\rho(z)|\boldsymbol{x} - \boldsymbol{z}|^{-1}(x^m - y^m)|\boldsymbol{x} - \boldsymbol{y}|^{-3}$$

$$- \sum_{A\neq B} \sum_{C\neq B,A} k^2c^4 M_A M_B M_C R_{CB}^{-1} R_{AB}^{-3} \boldsymbol{R}_{AB}^m$$

$$- \sum_{A\neq B} k^2c^4 M_A^2 M_B R_{AB}^{-4} \boldsymbol{R}_{AB}^m$$

$$- \sum_{A\neq B} k^2c^4 R_{AB}^{-3} \boldsymbol{R}_{AB}^m M_A \int_B d^3x \int_B d^3z \rho(x)\rho(z)|\boldsymbol{x} - \boldsymbol{z}|^{-1}. \tag{2.19}$$

Using the continuity equation and (1.2), og

$$\{\rho V^m cU_{,0}\}_B = \int_B d^3x \int_B d^3y\, kc^2 \rho V^m(x) \rho V^r(y)(x^r - y^r)|\boldsymbol{x} - \boldsymbol{y}|^{-3}$$

$$+ \sum_{A\neq B} kc^2 M_A M_B R_{AB}^{-3} \boldsymbol{R}_{AB}^r V_A^r V_B^m. \tag{2.20}$$

Since $g_m = -(1/2)(7\mathcal{V}_m + \mathcal{W}_m)$, eqs.(C12) and (C13) give

$$\{\rho c V^r(g_{r,m} - g_{m,r})\}_B =$$

$$- \int_B d^3x \int_B d^3y\, 4kc^2 \rho V^r(x) \rho V^m(y)(x^r - y^r)|\boldsymbol{x} - \boldsymbol{y}|^{-3}$$

$$+ \sum_{A\neq B} 4kc^2 M_A M_B R_{AB}^{-3}(\boldsymbol{R}_{AB}^m \boldsymbol{V}_A \cdot \boldsymbol{V}_B - V_A^m \boldsymbol{R}_{AB} \cdot \boldsymbol{V}_B). \tag{2.21}$$

A more precise evaluation of this term is given by (3.14).

The most complex term is

$$\{\rho g_{m,0}\}_B =$$

$$- \int_B d^3x \int_B d^3y\, 3k\rho(x)\rho V^m V^r(y)(x^r - y^r)|\boldsymbol{x} - \boldsymbol{y}|^{-3}$$

$$- \int_B d^3x \int_B d^3y\, 3k\rho(x)p(y)(x^m - y^m)|\boldsymbol{x} - \boldsymbol{y}|^{-3}$$

$$+ (1/2) \int_B d^3x \int_B d^3y\, k\rho(x)(x^m - y^m)$$

$$\Big\{\rho \boldsymbol{V} \cdot \boldsymbol{V}(y) - 3\rho(y)\big[\boldsymbol{V}(y) \cdot (\boldsymbol{x} - \boldsymbol{y})\big]^2 |\boldsymbol{x} - \boldsymbol{y}|^{-2}\Big\}|\boldsymbol{x} - \boldsymbol{y}|^{-3}$$

$$+\int_B d^3x \int_B d^3y \int_B d^3z\, k^2c^4\rho(x)\rho(y)\rho(z)|\boldsymbol{y}-\boldsymbol{z}|^{-3}(y^m-z^m)$$

$$\Big[(7/2)|\boldsymbol{x}-\boldsymbol{y}|^{-1}-(1/2)(y^s-x^s)(z^s-y^s)|\boldsymbol{x}-\boldsymbol{y}|^{-3}\Big]$$

$$-\sum_{A\neq B} 3kM_BR_{AB}^{-3}\big[\boldsymbol{R}_{AB}^r\{\rho V^rV^m\}_A+\boldsymbol{R}_{AB}^m\{p\}_A\big]$$

$$-(1/2)\sum_{A\neq B} kM_B\boldsymbol{R}_{AB}^mR_{AB}^{-3}\big[-\{\rho V^rV^r\}_A+3\boldsymbol{R}_{AB}^r\boldsymbol{R}_{AB}^sR_{AB}^{-2}\{\rho V^rV^s\}_A\big]$$

$$+(1/2)\sum_{A\neq B} k^2c^4M_A\boldsymbol{R}_{AB}^sR_{AB}^{-3}\int_B d^3x\int_B d^3y\,\rho(x)\rho(y)(x^s-y^s)(x^m-y^m)|\boldsymbol{x}-\boldsymbol{y}|^{-3}$$

$$+(1/2)\sum_{A\neq B}\sum_{C\neq A,B} k^2c^4M_AM_BM_CR_{AC}^{-3}R_{AB}^{-3}(\boldsymbol{R}_{AB}\cdot\boldsymbol{R}_{CA})\boldsymbol{R}_{AB}^m$$

$$-4\sum_{A\neq B}k^2c^4M_AM_B^2R_{AB}^{-4}\boldsymbol{R}_{AB}^m$$

$$+(7/2)\sum_{A\neq B}k^2c^4M_A\boldsymbol{R}_{AB}^mR_{AB}^{-3}\int_B d^3x\int_B d^3y\,\rho(x)\rho(y)|\boldsymbol{x}-\boldsymbol{y}|^{-1}$$

$$-(7/2)\sum_{A\neq B}\sum_{C\neq A,B}k^2c^4M_AM_BM_CR_{AC}^{-3}R_{AB}^{-1}\boldsymbol{R}_{AC}^m. \tag{2.22}$$

Proof of eq.(2.7)

Differentiating (2.5) and using the continuity equation V(4.2), we find that

$$DI_B^{rs}(t)=-\int_B(\rho V^n)_{,n}(x)(x^r-X_B^r)(x^s-X_B^s)\,d^3x$$

$$-\int_B\rho(x)\big[V_B^r(x^s-X_B^s)+(x^r-X_B^r)V_B^s\big]\,d^3x.$$

The second integral vanishes, from the definition of the centre of mass, and og

$$DI_B^{rs}(t)=\int_B\rho\big[V^r(x^s-X_B^s)+V^s(x^r-X_B^r)\big](x)\,d^3x. \tag{2.23}$$

Differentiating again and using (1.2) and the nonrelativistic Euler equation from (2.3), og

$$D^2I_B^{rs}(t)=\int_B\big[c(\rho V^r)_{,0}(x^s-X_B^s)+c(\rho V^s)_{,0}(x^r-X_B^r)\big](x)\,d^3x$$

$$-\int_B\rho(V^rV_B^s+V^sV_B^r)(x)\,d^3x$$

$$= 2\{\rho V^r V^s\}_B + 2\delta^{rs}\{p\}_B - 2M_B V_B^r V_B^s$$
$$- \sum_A kc^4 \int_B d^3x \int_A d^3y\, \rho(x)\rho(y)\Big[(x^r - y^r)(x^s - X_B^s)$$
$$+ (x^s - y^s)(x^r - X_B^r)\Big]|\boldsymbol{x} - \boldsymbol{y}|^{-3}. \qquad (2.24)$$

When $A \neq B$, one replaces $\boldsymbol{x} - \boldsymbol{y}$ by $\boldsymbol{R}_{AB}$, etc., in the usual manner, and shows that the contribution to the last term vanishes. When $A = B$, the terms proportional to X_B^r and X_B^s vanish. We use (3.1), and (2.7) follows.

3. Forces due to internal motions

The acceleration of the centre of mass as given by (2.9) agrees with Will's eq.(6.34) except for the terms involving $\{p\}$, $\{\rho\Pi\}$, I^{mn}, and β^{mn}. (In Will's equation, one must give the PPN parameters their standard values: $\beta = \gamma = 1$ and the rest zero.) We may suspect that some of the additional terms arise because of our special choice of mass density, and that they can be removed by a different choice. However, the β^{mn} terms cannot be explained in this way. They are a necessary consequence of the internal motions of the bodies (i.e they depend on $\boldsymbol{V} - \boldsymbol{V}_A$, the velocity of the matter in a body with respect to the centre of mass).

One can see on physical grounds that β_A^{mn} terms must be present: it would be irrational for the centre of mass motions to produce gravitational effects and the internal motions to produce none. To make this idea precise, we will show that (2.9), which is written in terms of $\boldsymbol{V}_A$ and $\boldsymbol{V} - \boldsymbol{V}_A$(the latter through the β_A^{mn}), can in fact be expressed in terms of $\boldsymbol{V}$ only.

From (1.2) og, in the Newtonian approximation,

$$\beta_B^{rs}(t) = \{\rho(V^r - V_B^r)(V^s - V_B^s)\}_B = \{\rho V^r V^s\}_B - M_B V_B^r V_B^s. \qquad (3.1)$$

In the first bracket of (2.9), we write $D^2 I_B^{mr}$ in terms of β_B^{mr} by (2.7), use (3.1) to eliminate the β_B^{mr}, and note that $M_A M_B V_A^r V_B^m = \{\rho V^r\}_A \{\rho V^m\}_B$. The contribution of the first bracket to (2.9) becomes

$$\sum_{A\neq B} kc^2 R_{AB}^{-3} \boldsymbol{R}^r_{AB}\Big[3M_B\{\rho V^m V^r\}_A + 4M_A\{\rho V^m V^r\}_B$$
$$- 3\{\rho V^r\}_A\{\rho V^m\}_B$$
$$- 4\{\rho V^m\}_A\{\rho V^r\}_B$$
$$+ \delta^{rm} M_A\{p\}_B - (1/2)M_A c^2\{\rho \mathcal{U}_{rmB}\}_B\Big]. \qquad (3.2)$$

Similarly, the second bracket contributes

$$\sum_{A\neq B} kc^2 R_{AB}^{-3} \boldsymbol{R}_{AB}^m$$

$$\Big[-2M_B\{\rho V^r V^r\}_A - M_A\{\rho V^r V^r\}_B + 4\{\rho V^r\}_A\{\rho V^r\}_B + M_B c^2\{\rho U_A\}_A$$

$$+ (1/2)M_A c^2\{\rho U_B\}_B - M_B c^2\{\rho\Pi\}_A + (3/2)M_B R_{AB}^{-2}\boldsymbol{R}_{AB}^r\boldsymbol{R}_{AB}^s\{\rho V^r V^s\}_A\Big]. \tag{3.3}$$

The expressions (3.2) and (3.3) are in terms of $\boldsymbol{V}$ only, as asserted.

In calculating $\boldsymbol{A}_{oth}$, we use the monopole approximation, which implies that we neglect terms like I_A^{mn}, eq.(2.5). But the smallness of the I_A^{mn} does not imply that the second derivatives $D^2 I_A^{mn}$ are always small. One must differentiate *before* making the monopole approximation, not afterwards; otherwise these terms will be missed. Very often, the $D^2 I_A^{mn}$ terms in (2.9) are indeed negligible. In calculating the Earth's orbital motion, for example, the slow precession of the axis of rotation can usually be ignored. In other cases, the $D^2 I_A^{mn}$ are quasiperiodic, and their time averages vanish over periods that are very small in comparison with the orbital periods of the bodies. One can then eliminate them by taking a time average of (2.9). However, one cannot neglect the $D^2 I_A^{mn}$ when bodies are tumbling or oscillating with periods comparable to their orbital periods.

Even when the $D^2 I_A^{mn}$ are negligible, the β_A^{mn} may not be. The commonest example is a rapidly rotating body whose axis of rotation changes only slowly.

Rotation and the Kepler problem

We now return to the original form (2.9) of the equations of motion, and try to understand the physical meaning of the internal-motion terms β_A^{mr}. They are obviously important for very massive and rapidly rotating bodies, but we need a more precise criterion for when they are significant.

For the moment, we shall not be concerned with the term $(3/2)kc^2 M_B R_{AB}^{-5}$ $\boldsymbol{R}_{AB}^m$ $\boldsymbol{R}_{AB}^r\boldsymbol{R}_{AB}^s\beta_A^{rs}$ in the second bracket of (2.9), which is just one among several small corrections to the inverse square law force (see Exercise 3.2). The other term involving β_A^{mr} is $3kc^2 R_{AB}^{-3}(M_B\beta_A^{mr} + M_A\beta_B^{mr})\boldsymbol{R}_{AB}^r$, from the first bracket. The best way to understand it is to consider it by itself, as a perturbation of the 2-body, Kepler problem (see Appendix F).

To simplify the calculations, we regard the bodies as being quasirigid, with angular velocities $\boldsymbol{\omega}_B$. One then has $\boldsymbol{V} - \boldsymbol{V}_B = \boldsymbol{\omega}_B \times (\boldsymbol{x} - \boldsymbol{X}_B)$, and $\beta_B^{mr} = \epsilon_{mnp}\epsilon_{rst}\omega_B^n\omega_B^s I_B^{pt}$ (we are being sloppy here – it would be more consistent to write $\boldsymbol{V}(x)$ instead of $\boldsymbol{V}$ and $\boldsymbol{\omega}_B(t)$ instead of $\boldsymbol{\omega}_B$). We take $A = 1$, $B = 2$, $\boldsymbol{R}_{AB} =$

$\boldsymbol{R}_{12} = \boldsymbol{X}_2 - \boldsymbol{X}_1 = \boldsymbol{R}$, and define a force $\boldsymbol{F}$ by

$$F^m = 3kc^2(M_2\beta_1^{mr} + M_1\beta_2^{mr})R^{-3}R^r. \tag{3.4}$$

We consider two bodies for which the equations of motion are

$$\begin{aligned} M_2\boldsymbol{A}_2 &= -kc^4M_1M_2R^{-3}\boldsymbol{R} + \boldsymbol{F}, \\ M_1\boldsymbol{A}_1 &= kc^4M_1M_2R^{-3}\boldsymbol{R} - \boldsymbol{F}, \end{aligned} \tag{3.5}$$

(in this Newtonian calculation we can drop the stars from M_1^* and M_2^*). The relative displacement $\boldsymbol{R}$ satisfies $mD^2\boldsymbol{R} = -mKR^{-3}\boldsymbol{R} + \boldsymbol{F}$, or

$$D^2\boldsymbol{R}^m = -KR^{-3}\boldsymbol{R}^m + B^{mr}R^{-3}\boldsymbol{R}^r, \tag{3.6}$$

where $K = kc^4(M_1 + M_2), B^{mr} = 3kc^2m^{-1}(M_2\beta_1^{mr} + M_1\beta_2^{mr}) = B^{rm}$, and $m = M_1M_2/(M_1 + M_2)$ is the reduced mass. We assume that the B^{mr} are constant, which is true if the bodies are cylindrically symmetric and rotate with constant angular velocities about their symmetry axes, for example.

We write $\boldsymbol{v} = D\boldsymbol{R}$ and $v = |\boldsymbol{v}|$. The energy per unit mass E, angular momentum per unit mass $\boldsymbol{L}$, and eccentricity vector $\boldsymbol{e}$ are defined by

$$E = (1/2)v^2 - KR^{-1}, \qquad \boldsymbol{L} = \boldsymbol{R} \times \boldsymbol{v}, \qquad \boldsymbol{e} = -K^{-1}\boldsymbol{L} \times \boldsymbol{v} - R^{-1}\boldsymbol{R}; \tag{3.7}$$

all three are constants of the motion when $\boldsymbol{F} = 0$. The eccentricity is defined to be $e = |\boldsymbol{e}|$, and from (3.7) og

$$e = (1 + 2EL^2/K^2)^{1/2}. \tag{3.8}$$

If one takes the dot product of the third of eqs.(3.7) with $\boldsymbol{R}$, and writes $\boldsymbol{e} \cdot \boldsymbol{R} = eR\cos\theta$ so that θ is the angle between $\boldsymbol{e}$ and $\boldsymbol{R}$, og $eR\cos\theta = K^{-1}L^2 - R$, or

$$R(1 + e\cos\theta) = \lambda, \tag{3.9}$$

where $\lambda = K^{-1}L^2$ is called the *semi-latus rectum* of the orbit, believe it or not. When e and λ are constant (which is true when $\boldsymbol{F} = 0$), eq.(3.9) is the equation of a conic section in terms of the polar coordinates R and θ. For elliptical orbits – the only case that we consider – one has $0 \leq e < 1$. It is convenient *not* to restrict the range of θ to $[0, \pi]$ (as one normally does in defining the angle between two vectors). We choose the sign of θ so that θ increases in the direction of the particle's motion, and find that

$$E = (1/2)[(RD\theta)^2 + (DR)^2] - KR^{-1}, \qquad L = R^2D\theta. \tag{3.10}$$

Exercise 3.1 Prove that $E, \boldsymbol{L}$, and $\boldsymbol{e}$ are constants of the motion if $\boldsymbol{F} = 0$. [For E, take the dot product of the equation of motion with $\boldsymbol{R}$; for $\boldsymbol{L}$, take the cross product with $\boldsymbol{R}$; for $\boldsymbol{e}$, take the cross product with $\boldsymbol{L}$ and note that $\boldsymbol{L}\times D\boldsymbol{v} = D(\boldsymbol{L}\times\boldsymbol{v}) = -KD(\boldsymbol{R}R^{-1})$.] ∎

When $\boldsymbol{F}\neq 0$, the quantities $E, \boldsymbol{L}$, and $\boldsymbol{e}$ are in general not constants of the motion. We define ΔE to be the change in E per orbit, and similarly for $\boldsymbol{L}$ and $\boldsymbol{e}$. It follows from the results of perturbation theory given in Appendix F that, for $\boldsymbol{F}$ given by (3.4),

$$\begin{aligned}\Delta E &= 0,\\ \Delta\boldsymbol{L} &= \pi LK^{-1}\Big[(1/2)B^{12}e^2\hat{L} + B^{23}(1+(1/4)e^2)\hat{e} - B^{31}(1+(3/4)e^2)\hat{\jmath}\Big],\\ \Delta\boldsymbol{e} &= -\pi K^{-1}e\Big[B^{23}\hat{L} + (1/2)B^{12}\hat{e} + (1/4)(B^{11}+B^{22})\hat{\jmath}\Big],\end{aligned} \tag{3.11}$$

where $\hat{L}$ and $\hat{e}$ are unit vectors in the directions of $\boldsymbol{L}$ and $\boldsymbol{e}$, respectively, and $\hat{\jmath} = \hat{L}\times\hat{e}$. We have neglected terms in e^3. The perihelion advance, in radians per orbit, is

$$\hat{\jmath}\cdot\Delta\boldsymbol{e} = -(1/4)\pi K^{-1}e(B^{11}+B^{22}). \tag{3.12}$$

To see what these results mean, we make some order of magnitude estimates. We suppose that $M_2 \ll M_1$, $\beta_2^{mr} = 0$, and $I_1^{mr} = I\delta^{mr}$, and we write $\omega_1^m = \omega^m$ and $I \approx (1/4)M_1r^2$, where r is the radius of body 1. We then have $B^{mr} \approx (3/4)kc^2M_1r^2(\omega^q\omega^q\delta^{mr} - \omega^m\omega^r)$. If ω^1 and ω^2 are small in comparison with ω^3, og $\omega^3 \approx \omega$, and $\boldsymbol{j}\cdot\Delta\boldsymbol{e} \approx -\,e(r\omega c^{-1})^2$. For the Sun, $(r\omega c^{-1})^2 \approx 4\times 10^{-11}$, and the contribution to the perihelion advance of Mercury ($e = 0.18$) is of order 10^{-3} seconds of arc per century – which is completely negligible. For the binary pulsar PSR 1913+16, with a rotational period of about 0.06 seconds, $r \approx 10$km, and $e \approx 0.6$, og $(r\omega c^{-1})^2 \approx 10^{-5}$, and the perihelion advance is about 0.3 degrees per year. This is a few percent of the relativistic perihelion advance given by III(6.6), which is approximately 4 degrees per year. For a very fast pulsar, with a rotational period of about a millisecond, the perihelion advance is of the order of a thousand times larger.

We conclude from this discussion that the rotation terms in (2.9) are likely to produce very small effects in the solar system but may be important elsewhere.

Exercise 3.2 Calculate ΔE, $\Delta\boldsymbol{L}$, and $\Delta\boldsymbol{e}$ for the case when $\boldsymbol{F}$ in (3.5) is given by $F^m = (3/2)kc^2M_BR_{AB}^{-5}\boldsymbol{R}^m_{AB}\boldsymbol{R}^r_{AB}\boldsymbol{R}^s_{AB}\beta_A^{rs}$ (the β_A^{rs} dependent term from the second bracket of (2.9)). Make rough estimates of the size of the effects to be expected, as was done for the force (3.4). ∎

Dipole contributions to the gravitational force

In calculating the force (2.9) we used the monopole approximation – the dipole, quadrupole, and higher multipole contributions were ignored. This is almost always justifiable, but occasionally the dipole terms do produce interesting, and even measurable effects.

We are going to calculate the term $\{V^r c^{-1}(g_{r,m} - g_{m,r})\}_B$ in (2.2) including the dipole contribution. Since $g_m = -(1/2)(7\mathcal{V}_m + \mathcal{W}_m) = -4\mathcal{V}_m + (1/2)(\mathcal{V}_m - \mathcal{W}_m)$, we find from (C11) that $g_{m,r} - g_{r,m} = -4(\mathcal{V}_{m,r} - \mathcal{V}_{r,m})$. As in (2.21), og

$$\begin{aligned}\{\rho c V^r(g_{r,m} - g_{m,r})\}_B = \\ - \int_B d^3x \int_B d^3y \, 4kc^2 \rho V^r(x) \rho V^m(y)(x^r - y^r)|\boldsymbol{x} - \boldsymbol{y}|^{-3} \\ + \sum_{A \neq B} \int_B d^3x \int_B d^3y \, 4kc^2 \rho V^r(x) \rho(y) \\ \Big[V^r(y)(x^m - y^m) - V^m(y)(x^r - y^r)\Big]|\boldsymbol{x} - \boldsymbol{y}|^{-3}.\end{aligned} \tag{3.13}$$

We use (1.9) to write $(x^m - y^m)|\boldsymbol{x} - \boldsymbol{y}|^{-3} = \Big[(x^m - X_A^m - (y^m - X_A^m)\Big]|\boldsymbol{x} - \boldsymbol{X}_A|^{-3}$ $\Big[1 + 3(\boldsymbol{x} - \boldsymbol{X}_A) \cdot (\boldsymbol{y} - \boldsymbol{X}_A)|\boldsymbol{x} - \boldsymbol{X}_A|^{-2}\Big]$, neglect terms quadratic in $y^m - X_A^m$, and find

$$\begin{aligned}\{\rho c V^r(g_{r,m} - g_{m,r})\}_B = \\ - \int_B d^3x \int_B d^3y \, 4kc^2 \rho V^r(x) \rho V^m(y)(x^r - y^r)|\boldsymbol{x} - \boldsymbol{y}|^{-3} \\ + 4kc^2 \sum_{A \neq B} \Big\{M_A M_B R_{AB}^{-3}(\boldsymbol{R}_{AB}^m \boldsymbol{V}_A \cdot \boldsymbol{V}_B - V_A^m \boldsymbol{R}_{AB} \cdot \boldsymbol{V}_B) \\ - M_B R_{AB}^{-3} V_B^r (S_A^{mr} - S_A^{rm}) \\ + 3 M_B R_{AB}^{-5} V_B^r \boldsymbol{R}_{AB}^q (\boldsymbol{R}_{AB}^m S_A^{qr} - \boldsymbol{R}_{AB}^r S_A^{qm})\Big\},\end{aligned} \tag{3.14}$$

where

$$\begin{aligned}S_A^{rs}(t) &= \int_A \rho(y)(y^r - X_A^r) V^s(y) \, d^3y \\ &= \int_A \rho(y)(y^r - X_A^r)(V^s - V_A^s) \, d^3y.\end{aligned} \tag{3.15}$$

It follows from (2.5) that $DI_A^{rs} = S_A^{rs} + S_A^{sr}$. The *spin angular momentum* or *spin* of body A is defined by $S^p_A = \epsilon_{pqr} S^{qr}_A$, or more explicitly by

$$\boldsymbol{S}_A(t) = \int_A \rho(y)(\boldsymbol{y} - \boldsymbol{X}_A) \times (\boldsymbol{V}(y) - \boldsymbol{V}_A) d^3y. \tag{3.16}$$

Og $\epsilon_{pmn} S^p_A = S^{mn}_A - S^{nm}_A$ and

$$S^{mn}_A = (1/2)\epsilon_{pmn} S^p_A + (1/2) D I^{mn}_A. \tag{3.17}$$

The spin-dependent terms of (3.14) can now be rewritten as

$$4kc^2 M_B R^{-3}_{AB} \sum_{A \neq B} \Big\{ - V^r_B (S^{mr}_A - S^{rm}_A) + 3R^{-2}_{AB} V^r_B \boldsymbol{R}^q_{AB} (\boldsymbol{R}^m_{AB} S^{qr}_A - \boldsymbol{R}^r_{AB} S^{qm}_A) \Big\}$$

$$\begin{aligned} =& 4kc^2 M_B R^{-3}_{AB} \sum_{A \neq B} \Big\{ - (\boldsymbol{V}_B \times \boldsymbol{S}_A)^m \\ &+ (3/2) R^{-2}_{AB} \Big[(\boldsymbol{S}_A \times \boldsymbol{R}_{AB}) \cdot \boldsymbol{V}_B \boldsymbol{R}^m_{AB} - (\boldsymbol{S}_A \times \boldsymbol{R}_{AB})^m \boldsymbol{V}_B \cdot \boldsymbol{R}_{AB} \\ &+ V^r_B \boldsymbol{R}^q_{AB} D I^{qr}_A \boldsymbol{R}^m_{AB} - (\boldsymbol{V}_B \cdot \boldsymbol{R}_{AB}) D I^{qm}_A \boldsymbol{R}^q_{AB} \Big] \Big\}. \end{aligned} \tag{3.18}$$

These terms are to be added to the right-hand side of (2.9). There are dipole terms in $g_{m,0}$ also, the calculation of which is left as an exercise.

Force due to a rotating body

We previously calculated, eq.(3.11), the perturbation of the motion of two bodies caused by the rotation of one of them. We now look more closely at the gravitational force produced by a rotating body – including the dipole contribution from (3.18). We again consider a system of two bodies with $M_2 \ll M_1$, and we suppose that the centre of mass of body 1 is at rest at the spatial origin, so that $\boldsymbol{V}_1 = 0$. The gravitational force $\boldsymbol{F}$ on body 2 is given by the right-hand side of (2.9) with $A = 1$ and $B = 2$, together with the Newtonian force $\boldsymbol{F}_{Newt}$ (cf. (2.4)) and the dipole terms – as usual, we drop the self-force terms. We assume that $\beta^{mr}_2 = 0$ and that we can neglect the terms in DI^{mn}_1, $D^2 I^{mn}_1$, $D^2 I^{mn}_2$, p, Π, and M^2_2. Writing $\boldsymbol{R}_{12} = \boldsymbol{R}$, $M_2 = M$, $\boldsymbol{V}_2 = \boldsymbol{V}$, og

$$\begin{aligned} F^m =& F^m_{\text{Newt}} + kc^2 M R^{-3} \Big\{ 4M_1 \boldsymbol{V} \cdot \boldsymbol{R} V^m - M_1 \boldsymbol{V} \cdot \boldsymbol{V} R^m + 3\beta^{mr}_1 \boldsymbol{R}^r \\ &+ (3/2) R^{-2} \boldsymbol{R}^m \boldsymbol{R}^r \boldsymbol{R}^s \beta^{rs}_1 + 4kc^4 R^{-1} \boldsymbol{R}^m M^2_1 - 4(\boldsymbol{V} \times S_1)^m \\ &+ 6R^{-2} \Big[(\boldsymbol{S}_1 \times \boldsymbol{R}) \cdot \boldsymbol{V} \boldsymbol{R}^m - (\boldsymbol{S}_1 \times \boldsymbol{R})^m \boldsymbol{V} \cdot \boldsymbol{R} \Big] \Big\}. \end{aligned} \tag{3.19}$$

Note that the terms in β^{mn}_1 fall off like R^{-2} as R increases, while the M^2_1 and $\boldsymbol{S}_1$ terms fall off like R^{-3}.

The spin-dependent part of $\boldsymbol{F}$ can be written as

$$kc^2MR^{-3}\Big[4\boldsymbol{S}_1 + 6R^{-2}(\boldsymbol{S}_1 \times \boldsymbol{R}) \times \boldsymbol{R}\Big] \times \boldsymbol{V}$$
$$= kc^2MR^{-3}\Big[-2\boldsymbol{S}_1 + 6R^{-2}(\boldsymbol{S}_1 \cdot \boldsymbol{R})\boldsymbol{R}\Big] \times \boldsymbol{V}. \tag{3.20}$$

To interpret (3.20), we recall a result of Newtonian mechanics. We suppose that the cartesian charts x and $\tilde{x}$ have the same spatial origin, and that x rotates with constant angular velocity $\boldsymbol{\Omega}$ in $\tilde{x}$ (so that $\tilde{x}$ rotates with angular velocity $-\boldsymbol{\Omega}$ in x). The force $\boldsymbol{f}$ on a particle of mass M in x is then related to the force $\tilde{\boldsymbol{f}}$ in $\tilde{x}$ by

$$\boldsymbol{f} = \tilde{\boldsymbol{f}} - 2M\boldsymbol{\Omega} \times \boldsymbol{V} - M\boldsymbol{\Omega} \times (\boldsymbol{\Omega} \times \boldsymbol{Z}), \tag{3.21}$$

where $\boldsymbol{V}$ and $\boldsymbol{Z}$ are the velocity and displacement from the origin of the particle in x. The term $-2M\boldsymbol{\Omega} \times \boldsymbol{V}$ is the *Coriolis force*, and $-M\boldsymbol{\Omega} \times (\boldsymbol{\Omega} \times \boldsymbol{Z})$ is the *centrifugal force*.

Eq.(3.20) has the form of a Coriolis force with $\boldsymbol{\Omega} = -\boldsymbol{\Omega}_{\text{LT}}$, where

$$\boldsymbol{\Omega}_{\text{LT}} = kc^2R^{-3}\Big[-\boldsymbol{S}_1 + 3R^{-2}(\boldsymbol{S}_1 \cdot \boldsymbol{R})\boldsymbol{R}\Big] \tag{3.22}$$

(cf. Soffel (3.3.73) - his $\boldsymbol{J}_\oplus$ is our $\boldsymbol{S}_1$). The subscripts LT commemorate Lense and Thirring, who discussed this effect in 1918. Eq.(3.20) is also of the same form as the magnetic force on a charged particle of velocity $\boldsymbol{V}$, and is sometimes called the *gravitomagnetic force*.

Exercise 3.3 There is nothing corresponding to a centrifugal force in (3.20). Calculate the quadrupole contributions to (3.13) and hence (3.19), and see whether they have the form of a centrifugal force. ∎

With some qualifications, we can take over the Newtonian result and interpret (3.20) as a Coriolis force that appears in the postnewtonian chart x because it is rotating with respect to a chart $\tilde{x}$. There is no Coriolis force in $\tilde{x}$: it is a "nonrotating frame" in the jargon of Newtonian mechanics. We have defined $\boldsymbol{\Omega}_{\text{LT}}$ so that it is the angular velocity of $\tilde{x}$ in x.

The situation is, however, more complex than in Newtonian mechanics because $\boldsymbol{\Omega}_{\text{LT}}$ is position dependent – we should really write it as $\boldsymbol{\Omega}_{\text{LT}}(\boldsymbol{R})$. There is no single, globally defined chart $\tilde{x}$, but a set of charts each defined on a small neighbourhood. To make this clearer, we define a chart $\tilde{x}(\boldsymbol{R}_0)$ in a neighbourhood of any point $\boldsymbol{R}_0$, with angular velocity $\boldsymbol{\Omega}_{\text{LT}}(\boldsymbol{R}_0)$ in x. In $\tilde{x}(\boldsymbol{R}_0)$ there is no Coriolis force on a particle at the point $\boldsymbol{R}_0$, and the Coriolis force is small when the particle is close to $\boldsymbol{R}_0$. One speaks of $\tilde{x}(\boldsymbol{R}_0)$ as a *local nonrotating frame* (or *chart*). One may visualize the local nonrotating frames as being pulled around by the spinning of body 1 (this is called *frame dragging*).

As a special case of (3.19), we take the spinning body 1 to be quasirigid and spherically symmetric with constant angular velocity $\boldsymbol{\omega}$. We write $I_1^{mr} = I\delta^{mr}$, where I is a constant (it is *half* the moment of inertia of the body about an axis through its centre). We then have $\boldsymbol{V} - \boldsymbol{V}_1 = \boldsymbol{\omega} \times (\boldsymbol{y} - \boldsymbol{X}_1)$, and $\beta_1^{mr} = I(\omega^q\omega^q\delta^{mr} - \omega^m\omega^r)$ from (2.6). Similarly from (3.16) og $S_1^m = I_1^{rr}\omega^m - I_1^{mr}\omega^r = 2I\omega^m$ or $\boldsymbol{S} = 2I\boldsymbol{\omega}$, and (3.19) becomes

$$\begin{aligned} F^m = F^m_{\text{Newt}} &+ kc^2MR^{-3}\Big\{(4M_1\boldsymbol{V}\cdot\boldsymbol{R}V^m - M_1\boldsymbol{V}\cdot\boldsymbol{V}\boldsymbol{R}^m + 4kc^4R^{-1}\boldsymbol{R}^mM_1^2 \\ &+ \Big[(9/2)I\boldsymbol{\omega}\cdot\boldsymbol{\omega} - (3/2)IR^{-2}(\boldsymbol{\omega}\cdot\boldsymbol{R})^2 + 12IR^{-2}(\boldsymbol{\omega}\times\boldsymbol{R})\cdot\boldsymbol{V}\Big]\boldsymbol{R}^m \\ &- 3I(\boldsymbol{\omega}\cdot\boldsymbol{R})\omega^m + 8I(\boldsymbol{\omega}\times\boldsymbol{V})^m - 12IR^{-2}(\boldsymbol{\omega}\times\boldsymbol{R})^m\boldsymbol{V}\cdot\boldsymbol{R}\Big\}. \end{aligned} \tag{3.23}$$

The motion of a gyroscope

The dipole forces due to a spinning body vanish at the point $\boldsymbol{R}_0$ in the local nonrotating frame $\tilde{x}(\boldsymbol{R}_0)$. A small body which is instantaneously at rest close to $\boldsymbol{R}_0$ in $\tilde{x}(\boldsymbol{R}_0)$ will experience no forces or torques, and will remain at rest in $\tilde{x}(\boldsymbol{R}_0)$. Since the angular velocity of $\tilde{x}(\boldsymbol{R}_0)$ in x is $\boldsymbol{\Omega}_{\text{LT}}$, it follows that the effect of the dipole forces is to make the second body precess with respect to the postnewtonian chart x with angular velocity $\boldsymbol{\Omega}_{\text{LT}}$. This is called the *Lense-Thirring precession*. A similar result holds if the second body spins about an axis whose direction is instantaneously fixed in $\tilde{x}(\boldsymbol{R}_0)$. The body experiences no torque in $\tilde{x}(\boldsymbol{R}_0)$, and its spin precesses with angular velocity $\boldsymbol{\Omega}_{\text{LT}}$ in $\tilde{x}(\boldsymbol{R}_0)$.

If one describes the motion of a spinning body (or *gyroscope*) in a postnewtonian chart, the Lense-Thirring precession must be regarded as a dynamical effect: the result of a torque produced by one body on the other (a "spin-spin interaction"). There is another effect – the *Thomas precession* – which is not due to external torques (it is *kinematical* rather than dynamical). The Thomas precession occurs even in special relativity, where the spin axis of a gyroscope whose centre of mass has velocity $\boldsymbol{V}$ and acceleration $\boldsymbol{A} = D\boldsymbol{V}$ in an inertial chart precesses with an angular velocity $\boldsymbol{\Omega}_{\text{Th}} = (1/2)c^{-2}\boldsymbol{A} \times \boldsymbol{V}$ (this is an approximate result – a more exact version is given later).

In a postnewtonian chart there is, in addition to the Lense-Thirring and Thomas precessions, a precession with angular velocity $\boldsymbol{\Omega}_{\text{geo}} = (3/2)\boldsymbol{V} \times \nabla U$ which is called the *geodetic precession*. One might argue from the occurrence of the ∇U, typical of Newtonian gravitational forces, that this is a dynamical effect. However, the geodetic and Thomas precessions can be derived by exactly similar arguments, so the distinction between dynamical and kinematical is perhaps dubious here. It is slightly odd that the Thomas precession was calculated in 1927, while the similar but more complex geodetic precession was discovered by de Sitter in 1916.

The Lense-Thirring, Thomas, and geodetic precessions are usually extremely small in comparison with the precessions due to non-gravitational torques and Newtonian gravitational torques. There is hope of measuring the Lense-Thirring and geodetic precessions for a gyroscope in orbit about the Earth. The quantal version of the Thomas precession is important in atomic and nuclear physics.

Exercise 3.4 Estimate the size of the Lense-Thirring and geodetic precesions for a gyroscope in a low orbit about the Earth. ■

Exercise 3.5 Write a short essay on the distinction between kinematical and dynamical quantities. ■

Calculation of the Thomas precession

We are going to calculate the special-relativistic Thomas precession, which is often omitted from textbooks. We shall then indicate how the proof can be modified to give the geodetic precession in postnewtonian theory. For more details, see Soffel, Section 3.3 and Appendix A.3.2.

We consider an inertial chart x in which the centre of mass of the gyroscope has 3 velocity $\boldsymbol{V}(t)$ and 3 acceleration $\boldsymbol{A}(t) = D\boldsymbol{V}(t)$ at time $t = x^0c^{-1}$. The path of the centre of mass is parameterized by the proper time τ (cf. III(8.4) and V(4.4)). The components in x of its 4 velocity $v(\tau)$ and 4 acceleration $a(\tau) = Dv(\tau)$ at proper time τ are given by $v(\tau) = (\gamma c, \gamma\boldsymbol{V})(t)$, $a(\tau) = (\gamma^4c^{-1}\boldsymbol{V}\cdot\boldsymbol{A}, \gamma^2\boldsymbol{A} + \gamma^4c^{-2}\boldsymbol{V}\cdot\boldsymbol{A}\boldsymbol{V})(t)$, where $\gamma = (1 - |\boldsymbol{V}|^2c^{-2})^{-1/2} = dt/d\tau$ and $d\gamma/dt = \gamma^3c^{-2}\boldsymbol{V}\cdot\boldsymbol{A}$. Note that we are following the usual custom of defining 4 vector fields to be functions of the proper time τ, while 3 vector fields are, quite often, functions of the coordinate time t. When both appear in the same equation, they are to be evaluated at the corresponding values of τ and t. To avoid ambiguity, derivatives may be written as $d/d\tau$ and d/dt, instead of D, where $d/d\tau = \gamma d/dt$.

We define an *instantaneous rest chart at* τ' to be an inertial chart $x_{(\tau')}$ in which the centre of mass of the gyroscope is at rest at the instant τ'. If the components in $x_{(\tau')}$ of any vector field b are $b^\mu_{(\tau')}$, one can write the defining condition for $x_{(\tau')}$ as $v^m_{(\tau')}(\tau') = 0$ (or $\boldsymbol{v}_{(\tau')}(\tau') = 0$), and it follows that $v^0_{(\tau')}(\tau') = c$.

We now define the *spin* to be a 4 vector field s which is a function of the proper time τ and is orthogonal to v. This means that $s \cdot v = \eta_{\mu\nu}s^\mu v^\nu = 0$, and hence that $Ds \cdot v + s \cdot a = 0$. We assume that $D\boldsymbol{s}_{(\tau')}(\tau') = 0$ for all τ' (i.e. no torque acts on the spin when the centre of mass is instantaneously at rest in an inertial chart). It follows that $Ds^0_{(\tau')}(\tau')c = s \cdot a(\tau')$, and that $Ds_{(\tau')}(\tau') = c^{-2}s \cdot av_{(\tau')}(\tau')$.

The last equation is Poincaré covariant, and therefore valid in any inertial chart at τ'. Since τ' is arbitrary, og

$$Ds = c^{-2} s \cdot a\, v \tag{3.24}$$

at all times in any inertial chart.

The equation $s \cdot v = 0$ is equivalent to $s^0 = c^{-1}\boldsymbol{s} \cdot \boldsymbol{V}$, and it follows from our previous expression for a that $s \cdot a = \gamma^2 \boldsymbol{s} \cdot \boldsymbol{A}$ and that (3.24) is equivalent to

$$\begin{aligned} ds^0/dt &= c^{-1}\gamma^2 \boldsymbol{s} \cdot \boldsymbol{A}, \\ d\boldsymbol{s}/dt &= c^{-2}\gamma^2 \boldsymbol{s} \cdot \boldsymbol{A}\, \boldsymbol{V} = c^{-2}\gamma^2\big[(\boldsymbol{A} \times \boldsymbol{V}) \times \boldsymbol{s} + (\boldsymbol{s} \cdot \boldsymbol{V})\boldsymbol{A}\big]. \end{aligned} \tag{3.25}$$

The expression for $d\boldsymbol{s}/dt$ in (3.25) is *not* the same as that quoted earlier for the Thomas precession: $d\boldsymbol{s}/dt = (1/2)c^{-2}(\boldsymbol{A} \times \boldsymbol{V}) \times \boldsymbol{s}$. It reduces to the earlier expression when $s^0 = c^{-1}\boldsymbol{s} \cdot \boldsymbol{V} = 0$.

The instantaneous rest chart $x_{(\tau')}$ is so far not uniquely defined (it is arbitrary to the extent of a spatial rotation, for example). We define a unique instantaneous rest chart of the gyroscope for each τ' by boosting the chart x with the velocity $\boldsymbol{V}(t')$, where t' is the value of the time coordinate of x that corresponds to τ'. The coordinates of $x_{(\tau')}$ are $x^{\mu}_{(\tau')} = S^{\mu}_{\nu}(t')x^{\nu}$, where the $S^{\mu}_{\nu}(t')$ are given by (B14) and (B15). The components of $s(\tau)$ in $x_{(\tau')}$ are $s^{\mu}_{(\tau')}(\tau) = S^{\mu}_{\nu}(t')s^{\nu}(\tau)$. When $\tau = \tau'$ (i.e. when the gyroscope is instantaneously at rest in $x_{(\tau')}$) og

$$\begin{aligned} s^0_{(\tau')}(\tau') &= 0, \\ s^m_{(\tau')}(\tau') &= s^m(\tau') - c^{-2}\gamma(\gamma+1)^{-1}V^m(t')\boldsymbol{s}(\tau') \cdot \boldsymbol{V}(t'). \end{aligned} \tag{3.26}$$

Dropping the primes, we define a derivative $D_{\mathrm{bf}x}$ (read *boost from* x) by

$$D_{\mathrm{bf}x} s^m(\tau) = \gamma^{-1} \lim{}_{\delta \to 0} \delta^{-1}\Big[s^m_{(\tau+\delta)}(\tau+\delta) - s^m_{(\tau)}(\tau)\Big]. \tag{3.27}$$

(This is a derivative with respect to t – hence the factor $\gamma^{-1} = d\tau/dt$.) Substituting from (3.26), og

$$\begin{aligned} D_{\mathrm{bf}x} s^m &= c^{-2}\gamma^2(\gamma+1)^{-1}(\boldsymbol{s} \cdot \boldsymbol{A})V^m \\ &\quad - c^{-4}\gamma^3(\gamma+1)^{-2}\boldsymbol{V} \cdot \boldsymbol{A})(\boldsymbol{s} \cdot \boldsymbol{V})V^m - c^{-2}\gamma(\gamma+1)^{-1}(\boldsymbol{s} \cdot \boldsymbol{V})A^m. \end{aligned} \tag{3.28}$$

Neglecting terms in $|\boldsymbol{V}|^2 c^{-2}$ gives

$$D_{\mathrm{bf}x}\boldsymbol{s} = (1/2)c^{-2}(\boldsymbol{A} \times \boldsymbol{V}) \times \boldsymbol{s}, \tag{3.29}$$

which is the the usual expression for the Thomas precession.

Eq.(3.29) is easy to interpret if the centre of mass of the gyroscope describes a closed orbit. The orbital period T in the chart x corresponds to a proper time interval T^*, and $x_{(\tau')} = x_{(\tau'+T*)}$ (the boosted chart at τ' is the same as that at $\tau' + T^*$). The change in the spin in one orbit, as measured in the chart $x_{(\tau')}$, is just the integral of (3.29) with respect to t around the orbit. A general interpretation of (3.29) in terms of astronomical observations is given by Soffel, Section 3.3.3.

The underlying, algebraic explanation for the Thomas precession is that the composition of two boosts is not in general a boost (i.e. the product of two symmetric matrices is not in general symmetric). If $B(\tau)$ is the boost that relates x and $x_{(\tau)}$, then the Lorentz transformation that relates $x_{(\tau)}$ and $x_{(\tau+\alpha)}$ is $B(\tau+\alpha) \circ B(\tau)^{-1}$. It is not a boost, but the product of a boost and a spatial rotation, and the spatial rotation gives rise to the Thomas precession. In the instantaneous rest chart at $\tau + \alpha$ that is found by boosting $x_{(\tau)}$, the spin has the same direction as in $x_{(\tau)}$, to first order in α.

It is straightforward to generalize the preceding argument to include the geodetic precession. One must replace inertial charts by postnewtonian charts, and the $\eta_{\mu\nu}$ by the components $g_{\mu\nu}$ of the postnewtonian metric. One writes $s \cdot v = g_{\mu\nu} s^\mu v^\nu = 0$, and differentiates to get $D_v s \cdot v + s \cdot a = 0$, where D_v is the covariant derivative along v and $a = D_v v$ (cf. III(8.9)).

Exercise 3.6 Derive the expression for the geodetic precession. ∎

4. Centre of mass and conservation of momentum

In the previous sections, we have defined the centre of mass in terms of the conserved mass density. Although this is easy to work with, and has the virtue of familiarity, it does have some disadvantages. There is no simple relationship between the velocity of the centre of mass and the momentum of a body, and there are small "self-acceleration" terms – contributions to the acceleration of a body that are due to the body itself – which are tedious to calculate.

One can avoid such difficulties by using a different definition of the centre of mass. The formulation of the theory is somewhat complicated, but the results are tidier. The reader must decide whether he wants to pick his way through more algebraic tangles. However, he should read at least the first few pages, which generalize some familiar ideas on the motion of the centre of mass of an isolated system. For those who love algebraic manipulation, Appendix E shows how the acceleration of the centre of mass is altered by a small, arbitrary change in the definition of the mass density.

It must be emphasized once more that we are seeking formal simplicity – equations that are elegant or easy to manipulate. There is, of course, no guarantee that the centre of mass which satisfies the simplest equations is also that which is most closely correlated with observations. The centre of a body's image as seen in a telescope may not correspond exactly to any of our definitions. The analysis of such questions is intricate and delicate, and will not be attempted here; we merely caution that the complications we sweep under our theoretical rug may reappear when one attempts to interpret observations.

Centre of mass of a system

The underlying idea of this section is that, instead of defining the centre of mass in terms of the conserved mass, we use the energy – which is also a conserved quantity. As before, we consider a system in which the matter is an ideal fluid with spatially bounded support. The mass at the instant $t = x^0 c^{-1}$ is now defined to be

$$M(t) = c^{-1} \int T^0(x)\, d^3x, \tag{4.1}$$

where $(x) = (x^0, \boldsymbol{x})$, and $T^0 = \rho^* c[1 - (1/2)U + \Pi + \Upsilon]$ as in VII(5.23). The centre of mass of the system at the instant t is defined to be the spatial point whose coordinates $X^m(t)$ satisfy

$$MX^m(t) = c^{-1} \int x^m T^0(x)\, d^3x. \tag{4.2}$$

One shows from VII(5.14) and VII(5.8) that

$$M = c^{-1}P^0 + \rho L^3 O_4, \qquad DM = dM/dt = \rho c L^2 O_5, \tag{4.3}$$

where cP^0 is the energy of the system. The velocity of the centre of mass of the system is $\boldsymbol{V} = D\boldsymbol{X}$, and (4.2) and (4.3) give

$$MV^m(t) = \int x^m T^0{}_{,0}(x)\, d^3x + \rho c L^3 O_5. \tag{4.4}$$

Substituting from VII(5.25) and using the divergence theorem, og

$$MV^m(t) = \int \big\{T^m + (1/2)cx^m(\rho_{,0}U - \rho U_{,0}) - (1/2)\rho c(\mathcal{V}_m - \mathcal{W}_m)\big\}(x)\, d^3x + \rho c L^3 O_5. \tag{4.5}$$

We get rid of the $\rho_{,0}$ terms by using the continuity equation in the usual way. We show from VII(5.16) and the definition of $\mathcal{W}_m$ that

$$\int x^m(\rho_{,0}U - \rho U_{,0})(x)\, d^3x = \int \rho(\mathcal{V}_m - \mathcal{W}_m)(x)\, d^3x + \rho L^3 O_5, \tag{4.6}$$

and hence from VII(5.23) and VII(5.15)

$$MV^m(t) = \int T^m(x)\, d^3x + \rho c L^3 O_5 = P^m(t) + \rho c L^3 O_5. \tag{4.7}$$

Since $DP^m = \rho c^2 L^2 O_6$ from VII(5.22), og $DV^m = c^2 L^{-1} O_6$. The velocity of the centre of mass of the system is therefore constant to postnewtonian order (that is, $\boldsymbol{V} = \text{constant} + cO_5$).

Definitions of mass

We next consider a system that consists of a discrete set of bodies. The mass m'_B and the centre of mass $\boldsymbol{X}'_B$ of body B are defined by equations exactly like (4.1) and (4.2), except that the integrations are over the spatial region occupied by B:

$$m'_B(t) = c^{-1} \int_B T^0(x)\, d^3x, \tag{4.8}$$

$$m'_B X'^m_B(t) = c^{-1} \int_B x^m T^0(x)\, d^3x. \tag{4.9}$$

The alert reader will have deduced from the presence of the primes that these are provisional definitions, to be improved in the light of experience. The mass of the system (which we may now call the *total mass*) is $M = \sum_B m'_B$. Note that in this section we are not using the summation convention for upper-case Latin indices: summation over particles will be shown explicitly.

The mass m'_B is in general not conserved. If we differentiate (4.8) and note that ρ vanishes on the boundary of B, we find from VII(5.25) that

$$\begin{aligned} Dm'_B(t) &= \int_B T^0{}_{,0}(x)\, d^3x \\ &= (1/2)c\{\rho_{,0} U - \rho U_{,0}\}_B + \rho c L^2 O_5, \end{aligned} \tag{4.10}$$

where we again use the notation $\{f\}_B = \int_B f(x)\, d^3x$, for any function f.

There is no difficulty in formulating the theory with nonconstant m'_B: we are accustomed to dealing with nonconstant masses even in special relativity. A practical objection is that m'_B depends on U, the potential due to the whole mass distribution. Consequently, m'_B is of little use in problems where the aim is to calculate the mass distribution – since one cannot determine m'_B until the mass distribution is already known. It is arguable that one should ignore this objection, formulate the theory in terms of the m'_B, and worry later about the practicalities of calculation. However, we shall meekly follow the conventional path and introduce a more complicated but approximately constant mass.

In the definition of mass (4.8), we replace U by U_B in T^0, where U_B is defined by an equation like (1.5):

$$U_B(x) = kc^2 \int_B \rho(y)|\boldsymbol{x} - \boldsymbol{y}|^{-1} d^3y, \tag{4.11}$$

and $(y) = (x^0, \boldsymbol{y})$. We also replace Υ by $\Upsilon_B = (1/2)c^{-2}|\boldsymbol{V} - \boldsymbol{V}_B|^2$ – which means that we consider only internal motions of the body. The new mass m_B of the body is defined by (cf. (2.6))

$$\begin{aligned} m_B &= \left\{\rho^*(1 + (1/2)c^{-2}|\boldsymbol{V} - \boldsymbol{V}_B|^2 - (1/2)U_B + \Pi)\right\}_B \\ &= M_B^* + (1/2)c^{-2}\beta_B^{rr} - (1/2)\{\rho U_B\}_B + \{\rho\Pi\}_B. \end{aligned} \tag{4.12}$$

It is permissible in this equation to regard $\boldsymbol{V}_B$ as the velocity of the Newtonian centre of mass, defined by $\boldsymbol{M}_B^* \boldsymbol{V}_B = \{\rho^* \boldsymbol{V}\}_B$ (the error in m_B is of order $\rho L^3 O_4$). As usual, we have dropped the stars in the postnewtonian terms.

By means of VII(4.13) and (2.3), we calculate the derivative of m_B:

$$Dm_B(t) = \sum_{A \neq B} \left\{\rho U_{A,m}(V^m - V_B^m)\right\}_B + \rho c L^2 O_5. \tag{4.13}$$

If *in addition* we make the approximation that $\boldsymbol{x} - \boldsymbol{y} = \boldsymbol{X}_B - \boldsymbol{X}_A = \boldsymbol{R}_{AB}$ when $\boldsymbol{x}$ is in B and $\boldsymbol{y}$ in A, og

$$\begin{aligned} Dm_B(t) &= \sum_{A \neq B} -kc^2 \boldsymbol{R}_{AB}^m R_{AB}^{-3} \{\rho(V^m - V_B^m)\}_B + \rho c L^2 O_5 \\ &= \rho c L^2 O_5. \end{aligned} \tag{4.14}$$

The approximation amounts to ignoring the variation in the gravitational forces across the bodies – the tidal forces. This is not permissible, and m_B will not be constant, if the tidal forces are large. The mass m_B is sometimes called the *(postnewtonian) inertial mass*.

There is a neater way to formulate the approximation required for the derivation of (4.14). We recall that in the Newtonian approximation, which is adequate here, $\rho c^2 U_{,m} = p_{,m} + \rho D_t V^m + \rho c^2 L^{-1} O_4$. We *assume* that inside body B

$$\rho c^2 U_{B,m} = p_{,m} + \rho D_t(V^m - V_B^m) + \rho c^2 L^{-1} O_4, \tag{4.15}$$

where $\boldsymbol{V}_B$ is again the velocity of the Newtonian centre of mass (as usual, we can write the density as ρ or ρ^* in the Newtonian approximation). Eq.(4.15) says that the pressure in B depends only on the potential due to B and the internal motion (the velocity with respect to the centre of mass). We can expect the approximation to be valid when the tidal forces due to other bodies are small in comparison with the forces due to B itself – it may not hold for close binary stars, for example.

It follows from (4.15) that inside B one has

$$\rho c^2(U_{,m} - U_{B,m}) = \rho D V_B^m + \rho c^2 L^{-1} O_4. \tag{4.16}$$

Substituting in (4.13), og $Dm_B(t) = \left\{\rho(U_{,m} - U_{B,m})(V^m - V_B^m)\right\}_B + \rho c L^2 O_5 = \left\{\rho c^{-2} D V_B^m (V^m - V_B^m)\right\}_B + \rho c L^2 O_5 = \rho c L^2 O_5$ – which is (4.14).

Velocity of the centre of mass

It is convenient to define the conserved density ρ_B^* of body B by $\rho_B^*(x) = \rho^*(x)$ inside B, and $\rho_B^*(x) = 0$ outside B. The invariant density ρ_B is defined in an exactly similar way. Since we are assuming that the mass distribution consists of discrete, nonoverlapping bodies, ρ_B^* obeys the same continuity equation as ρ^*. The reason for introducing the ρ_B^* is that several of our previous results are little changed when ρ^* is replaced by ρ_B^*.

We begin with a generalization of VII(5.25). In analogy with VII(5.23), we define T_B^0 and T_B^m by

$$\begin{aligned} T_B^0 &= c\rho_B^*(1 - (1/2)U_B + \Pi + \Upsilon_B), \\ T_B^m &= \rho_B^* V^m (1 - U_B + \Pi + \Upsilon_B + p/\rho_B^* c^2) + (1/2)\rho_B^* c(\mathcal{V}_{mB} - \mathcal{W}_{mB}). \end{aligned} \tag{4.17}$$

where $\Upsilon_B = (1/2)c^{-2}|\boldsymbol{V} - \boldsymbol{V}_B|^2$, and $\mathcal{V}_{mB}$ and $\mathcal{W}_{mB}$ are defined in the same way as $\mathcal{V}_m$ and $\mathcal{W}_m$, except that the range of integration is restricted to B (cf. (4.11)). In small, postnewtonian terms we may write ρ_B in place of ρ_B^*. Following the proof of VII(5.25) almost exactly, we find that

$$\begin{aligned} T_{B,0}^0 = &-T_{B,m}^m + (1/2)c(\rho_{B,0}U_B - \rho_B U_{B,0}) \\ &+ (1/2)[\rho_B c(\mathcal{V}_{mB} - \mathcal{W}_{mB})]_{,m} + \rho_B V^m (U - U_B)_{,m} \\ &+ c^{-2} p_{,m} V_B^m - \rho U_{,m} V_B^m - \rho c^{-2}(\boldsymbol{V} - \boldsymbol{V}_B)\cdot \boldsymbol{A}_B + \rho c L^{-1} O_5. \end{aligned} \tag{4.18}$$

The centre of mass $\boldsymbol{X}_B$ of B is defined by

$$m_B X_B^m(t) = c^{-1}\{x^m T_B^0\}. \tag{4.19}$$

Note that we are using the same symbol $\boldsymbol{X}_B$ as for the Newtonian centre of mass and for the postnewtonian centre of mass defined in the last section. If the reader finds this confusing, he should use different symbols, and similarly for the velocity $\boldsymbol{V}_B$ and acceleration $\boldsymbol{A}_B$ defined below.

Differentiating (4.19) with respect to t and using (4.18) og

$$D(m_B X_B^m)(t) = \{x^m T^0_{B,0}\}$$

$$= \{T_B^m - (1/2)\rho c(\mathcal{V}_{mB} - \mathcal{W}_{mB}) + (1/2)c x^m(-\rho U_{B,0} + \rho_{,0}U_B)$$

$$+ x^m \rho V^r(U_{,r} - U_{B,r}) + x^m V_B^r(c^{-2}p_{,r} - \rho U_{,r} + \rho c^{-2}A_B^r)$$

$$- x^m \rho c^{-2} V^r A_B^r\}_B + \rho c L^3 O_5. \qquad (4.20)$$

We note that $\{x^m p_{,r}\}_B = -\delta_r^m \{p\}_B$, and that $\{x^m \rho U_{B,r}\}_B = -(1/2)\{\rho \mathcal{U}_{mrB}\}_B$, where (cf. (C8))

$$\mathcal{U}_{mrB}(x) = kc^2 \int_B d^3y \rho(y)(x^m - y^m)(x^r - y^r)|\boldsymbol{x} - \boldsymbol{y}|^{-3}. \qquad (4.21)$$

In the same way that we proved VII(5.16) and (4.6), we show that

$$\{\rho V^m U_B\}_B = \{\rho c \mathcal{V}_{mB}\}_B + \rho c L^3 O_5, \qquad (4.22)$$

$$\{x^m(\rho_{,0}U_B - \rho U_{B,0})\}_B = \{\rho(\mathcal{V}_{mB} - \mathcal{W}_{mB})\}_B + \rho L^3 O_5. \qquad (4.23)$$

Eq.(4.20) now becomes

$$D(m_B X_B^m)(t) = \{T_B^m\}_B + \left\{x^m \rho(V^r - V_B^r)(U_{,r} - U_{B,r}) - x^m \rho c^{-2}(V^r - V_B^r)A_B^r\right\}_B$$

$$- c^{-2}\{p\}_B V_B^m + (1/2)\{\rho \mathcal{U}_{mrB}\}_B V_B^r + \rho c L^3 O_5. \qquad (4.24)$$

We have not used (4.14) or (4.16) in the derivation of (4.24). If we do use them, we find that

$$m_B V_B^m(t) = \{T_B^m\}_B - c^{-2}\{p\}_B V_B^m + (1/2)\{\rho \mathcal{U}_{mrB}\}_B V_B^r + \rho c L^3 O_5. \qquad (4.25)$$

Writing T_B^m in terms of T_B^0 from (4.17) and using (4.22), og

$$m_B V_B^m(t) = \left\{T_B^0 c^{-1} V^m + p c^{-2}(V^m - V_B^m)\right\}_B$$

$$- (1/2)\{\rho c \mathcal{W}_{mB}\}_B + (1/2)V_B^r\{\rho \mathcal{U}_{mrB}\}_B + \rho c L^3 O_5. \quad (4.26)$$

This equation is equivalent to Will's eq.(6.26). To prove it, note that the last two terms can be written as

$$-\{(1/2)\rho c \mathcal{W}_{mB}\}_B + (1/2)V_B^r\{\rho \mathcal{U}_{mrB}\}_B$$

$$= -(1/2)kc^2 \int_B d^3x \int_B d^3y \left[\rho(x)\rho(y)(x^m - y^m)(V^r(y) - V_B^r)(x^r - y^r)|\boldsymbol{x} - \boldsymbol{y}|^{-3}\right]$$

$$+ \rho c L^3 O_5. \qquad (4.27)$$

Acceleration of the centre of mass

To calculate the acceleration $\boldsymbol{A}_B = D\boldsymbol{V}_B$ of the centre of mass, we differentiate (4.25) and again use (4.14):

$$m_B A_B^m(t) = D\{T_B^m\}_B - c^{-2} D\Big\{p\delta_{mr} - (1/2)c^2\rho\mathcal{U}_{mrB}\Big\}_B V_B^r$$
$$- c^{-2}\Big\{p\delta_{mr} - (1/2)c^2\rho\mathcal{U}_{mrB}\Big\}_B A_B^r + \rho c^2 L^2 O_6. \qquad (4.28)$$

The first term can be evaluated from VII(5.20). Since

$$\rho^* D_t\Big[V^m(1 + 3U + \Pi + \Upsilon + p/\rho c^2) + cg_m\Big] =$$
$$\rho^* D_t\Big[V^m(1 - U_B + \Pi + \Upsilon_B + p/\rho c^2) + (1/2)c(\mathcal{V}_{mB} - \mathcal{W}_{mB})\Big]$$
$$+ \rho^* D_t\Big[V^m(3U + U_B + \Upsilon - \Upsilon_B) - (1/2)c(\mathcal{V}_{mB} - \mathcal{W}_{mB}) - (1/2)c(7\mathcal{V}_m + \mathcal{W}_m)\Big],$$

og

$$\Big\{\rho^* D_t\Big[V^m(1 + 3U + \Pi + \Upsilon + p/\rho c^2) + cg_m\Big]\Big\}_B$$
$$= D\{T_B^m\}_B + D\Big\{\rho^*\Big[V^m(3U + U_B + \Upsilon - \Upsilon_B) - 4c\mathcal{V}_m$$
$$+ (1/2)c(\mathcal{V}_m - \mathcal{V}_{mB}) - (1/2)c(\mathcal{W}_m - \mathcal{W}_{mB})\Big]\Big\}_B$$
$$= D\{T_B^m\}_B + 4D\{\rho^*(V^m U - c\mathcal{V}_m)\}_B$$
$$+ D\Big\{\rho^*\Big[V^m(U_B - U) + (1/2)c(\mathcal{V}_m - \mathcal{V}_{mB}) - (1/2)c(\mathcal{W}_m - \mathcal{W}_{mB})\Big]\Big\}_B$$
$$+ c^{-2} D\beta_B^{mr} V_B^r + c^{-2}(\beta_B^{mr} + M_B V_B^r V_B^m) A_B^r$$
$$+ (1/2)c^{-2} M_B V_B^r V_B^r A_B^m. \qquad (4.29)$$

In the last step we have used the equation

$$\{\rho V^m(\Upsilon - \Upsilon_B)\}_B = c^{-2}\beta_B^{mr} V_B^r + (1/2)c^{-2} M_B V_B^r V_B^r V_B^m, \qquad (4.30)$$

where the β_B^{mr} are given by (3.1). The M_B can of course be replaced by m_B.

The expression $\{\rho^*(V^m U_A - c\mathcal{V}_{mA})\}_B$ vanishes when $A = B$, from (4.22). When $A \neq B$, we use the monopole approximation and find that

$$\{\rho^*(V^m U - c\mathcal{V}_m)\}_B = \sum_{A \neq B} kc^2 M_A M_B R_{AB}^{-1}(V_B^m - V_A^m) + \rho c L^3 O_5,$$

and hence

$$D\{\rho^*(V^m U - c\mathcal{V}_m)\}_B =$$

$$\sum_{A\neq B} kc^2 M_A M_B\big[R_{AB}^{-1}(A_B^m - A_A^m) - R_{AB}^{-3}\boldsymbol{R}^r_{AB}(V_B^r - V_A^r)(V_B^m - V_A^m)\big]$$

$$+ \rho c^2 L^2 O_6. \qquad (4.31)$$

One substitutes this expression into (4.29) and uses the Newtonian, monopole approximations for A_A^m and A_B^m.

We integrate VII(5.20) over the interior of body B, and use (4.29) and the following equations (cf. the proof of VII(5.21)):

$$\begin{gathered}\{\rho U_{B,m}\}_B = 0, \qquad \{pU_{B,m} + \rho c^2\Phi_{4B,m}\}_B = 0,\\ \{\rho(2\Upsilon U_{B,m} + \Phi_{1B,m})\}_B = 0, \qquad \{\rho(U_B U_{B,m} + \Phi_{2BB,m})\}_B = 0,\\ \{\rho(\Pi U_{B,m} + \Phi_{3B,m})\}_B = 0, \qquad \{\rho V^r g_{rB,m}\}_B = 0,\end{gathered} \qquad (4.32)$$

where the potentials Φ_{2A} and Φ_{2AB} are defined by

$$\begin{aligned}\Phi_{2A}(x) &= kc^2\int_A \rho U(y)|\boldsymbol{x}-\boldsymbol{y}|^{-1}\,d^3y,\\ \Phi_{2AB}(x) &= kc^2\int_A \rho U_B(y)|\boldsymbol{x}-\boldsymbol{y}|^{-1}\,d^3y\\ &= k^2c^4\int_A d^3y\int_B d^3z\rho(y)\rho(z)|\boldsymbol{x}-\boldsymbol{y}|^{-1}|\boldsymbol{y}-\boldsymbol{z}|^{-1}.\end{aligned} \qquad (4.33)$$

We find that

$$\begin{aligned}D\{T_B^m\}_B &+ D\sum_{A\neq B}\big\{\rho^*\big[-V^m U_A + (1/2)c\mathcal{V}_{mA} - (1/2)c\mathcal{W}_{mA}\big]\big\}_B\\ &+ c^{-2}D\beta_B^{mr}V_B^r + c^{-2}(\beta_B^{mr} + M_B V_B^r V_B^m)A_B^r\\ &+ (1/2)c^{-2}M_B V_B^r V_B^r A_B^m + 4D\big\{\rho^*(V^m U - c\mathcal{V}_m)\big\}_B\\ &- \sum_{A\neq B}\big\{\rho c^2 U_{A,m} + 3(pU_{A,m} + \rho c^2\Phi_{4A,m})\\ &+ \rho c^2(4\Upsilon U_{A,m} + 2\Phi_{1A,m}) + \rho c^2(\Pi U_{,Am} + \Phi_{3A,m}) + \rho c V^r g_{rA,m}\big\}_B\\ &- 2\{\rho c^2(UU_{,m} + \Phi_{2,m})\}_B + 2\{\rho c^2(U_B U_{B,m} + \Phi_{2BB,m})\}_B\\ = \rho c^2 L^2 O_6.\end{aligned} \qquad (4.34)$$

Substituting this into (4.28) and using (2.7), og

$$
\begin{aligned}
&m_B A_B^m(t) - D \sum_{A\neq B} \left\{\rho^*\left[V^m U_A - (1/2)c\mathcal{V}_{mA} + (1/2)c\mathcal{W}_{mA}\right]\right\}_B \\
&- (1/2)c^{-2}D^3 I_B^{mr} V_B^r - (1/2)c^{-2}D^2 I_B^{mr} A_B^r - c^{-2}M_B \boldsymbol{V}_B \cdot \boldsymbol{A}_B V_B^m \\
&- (1/2)c^{-2}M_B \boldsymbol{V}_B \cdot \boldsymbol{V}_B A_B^m - 4D\{\rho^*(V^m U - c\mathcal{V}_m)\}_B \\
&+ \sum_{A\neq B} \left\{\rho c^2 U_{A,m} + 3(pU_{A,m} + \rho c^2 \Phi_{4A,m}) + \rho c^2(4\Upsilon U_{A,m} + 2\Phi_{1A,m})\right. \\
&\quad \left. + \rho c^2(\Pi U_{A,m} + \Phi_{3A,m}) + \rho c V^r g_{rA,m}\right\}_B \\
&+ 2\{\rho c^2(UU_{,m} + \Phi_{2,m})\}_B - 2\{\rho c^2(U_B U_{B,m} + \Phi_{2BB,m})\}_B + \rho c^2 L^2 O_6.
\end{aligned}
\tag{4.35}
$$

Eq.(4.35) has been much more difficult to derive than the analogous equation (2.2) for the conserved mass density. The advantage of (4.35) is that it does not contain self-acceleration terms. On the other hand, its derivation assumes the constancy of the mass m_B; the results are suspect if tidal forces are large.

One might think that the easiest way to calculate the acceleration would be simply to compare (4.35) and (2.2) and see which terms differ. However, care is needed (see the paragraph following (2.4)). It is probably safer to evaluate (4.35) directly by using the equations at the end of Section 2.

Exercise 4.1 Calculate $\boldsymbol{A}_B$ from (4.35). [Note that ρ is the invariant mass density, and U is defined in terms of ρ. The Newtonian monopole term in the acceleration is therefore expressed in terms of the invariant masses M_A, which are not constant in time.] ∎

Exercise 4.2 Express the result of Exercise 4.1 in terms of the conserved masses M_A^*, and compare your result with (2.2). [Use VII(4.8). Neglect the difference between ρ and ρ^* in the Newtonian quadrupole term.] ∎

Exercise 4.3 Express the result of Exercise 4.1 in terms of the inertial masses m_A defined by (4.12), and compare your result with that of Exercise 4.2 and with (2.4) and (2.9). ∎

Gravitational mass

The mass m_B that we have used in this chapter is the postnewtonian inertial mass of body B. From (4.12) and (2.8) og

$$m_B = M_B^* + (1/2)c^{-2}\beta_B^{rr} - (1/2)\{\rho U_B\}_B + \{\rho\Pi\}_B.$$
$$= M_B^* + (3/2)c^{-2}\beta_B^{rr} + 3c^{-2}\{p\}_B - \{\rho U_B\}_B + \{\rho\Pi\}_B - (1/2)c^{-2}D^2 I_B^{rr}. \tag{4.36}$$

The last equation is almost the same as that for the gravitational mass $\mathcal{M}_B$ introduced in Chapter V. To see this, we use VII(4.8) to rewrite the expression V(2.6) for $\mathcal{M}_B$ in terms of ρ^*:

$$\mathcal{M}_B = M_B^* + 3\{\rho\Upsilon\}_B + 3c^{-2}\{p\}_B - \{\rho U\}_B + \{\rho\Pi\}_B. \tag{4.37}$$

If we replace U by U_B and $\boldsymbol{V}$ by $\boldsymbol{V} - \boldsymbol{V}_B$ and subtract $(1/2)c^{-2}D^2 I_B^{rr}$, eq.(4.37) becomes (4.36). The $D^2 I_B^{rr}$ term is usually negligible – it vanishes for a quasirigid body. Thus m_B differs from the gravitational mass $\mathcal{M}_B$ in omitting the potentials due to other bodies and the effects of the centre of mass motion.

Unless the reader is going to be a very serious calculator of postnewtonian motions (a fate similar to being a chartered accountant), she need not be much concerned with the different ways of defining the centre of mass. The calculated values of observable quantities are usually insensitive to the choice; the most notable exception is the Nordtvedt effect (see Section 2). The main purpose of our discussion is cautionary: to show the complexities and ambiguities that arise when one takes even a small step from the simplicities of the Newtonian theory.

This has been an arduous chapter. The postnewtonian theory is extremely important because it enables us to calculate observable quantities. However, it is hard to love. We have done our duty, and need feel no guilt in abandoning it for more diverting company.

IX Exact Theories of Gravity

1. Limitations of the postnewtonian theory

The postnewtonian theory accounts quite well for almost all the gravitational phenomena that we observe at present. However, there are a few observations that it cannot explain, and we may expect that there will soon be many more. For example, the theory is unreliable for calculating the properties of a neutron star, in which the dimensionless potential may have a value of 0.2 or more. The postnewtonian theory cannot account for gravitational waves. In such a wave, the space and time derivatives are of the same order of magnitude[1] – which contradicts the basic postnewtonian assumption that the time derivatives be of higher postnewtonian order than the space derivatives. Neither can the theory deal with the early universe, where the density of matter and radiation is very great. In all these cases, a more exact theory is required.

Can we find an exact theory by generalizing the postnewtonian theory in a unique, or at least a plausible manner? We might be encouraged by previous successes in generalizing the Newtonian theory and gravitostatics – but what we are now attempting is fundamentally different. There we were guided by well established hypotheses; we had only to refine them a little and elicit their further consequences. Although there was no guarantee of success – a hypothesis stretched too far might have broken – there was a unique, obvious path to be followed.

Beyond the postnewtonian theory there is no unique path. We are faced with real choices which lead to quite different theories. We have already had intimations of this in Chapter VII, Sections 6 and 7, where we wrote the postnewtonian field equations in several different forms. The equations are covariant under semipoincaré transformations (sometimes combined with a gauge transformation), and it is natural to try to generalize them so that they are covariant under arbitrary Poincaré transformations, or possibly under arbitrary, smooth coordinate transformations. The generalization is ambiguous for two reasons (which are not completely independent). First, one must choose the field variables that are to describe the gravitational field: the n_μ, or $q_m = -n_m/n_0$ and $\beta = \alpha n^2$, or q_m and ω, or some combinations of the $g_{\mu\nu}$. Second, one has to guess terms that do not appear in the postnewtonian equations, but which may nevertheless be important. For example, a postnewtonian field equation may contain a term proportional to $\phi_{,m}\phi_{,m}$(which is O_4 if ϕ is O_2), but not one proportional to $(\phi_{,0})^2$(which is O_6). In a gravitational wave the terms may be comparable in size, and one may have to add a $(\phi_{,0})^2$ term to the field equation. With luck, one can deduce the extra term from considerations of covariance, but this is not always possible.

In the next few sections we consider just two of the many relativistic theories that have the correct postnewtonian limit. Each is plausible, from a particular point of view, but there is no way to decide their correctness except by further experiments.

2. The Einstein theory

The Einstein theory was described very briefly in Chapter III. We now give a short account of the Einstein field equations, and show that they do have the correct postnewtonian limit. We then consider their plausibility as generalizations of the postnewtonian equations. A modern but not excessively formal account of the Einstein theory is given in [Wald 84]; a more elementary treatment is in [Schutz 85]. Several older books, which are still well worth reading, are included in the References.

We write the components of the spacetime metric in a chart x as $g_{\mu\nu}$ and their partial derivatives with respect to x^π as $g_{\mu\nu,\pi}$, in the usual manner. The Christoffel symbols of the first and second kinds are defined by III(8.8) or (D15). If u, v, and w are vector fields with components u^μ, v^μ, and w^μ in x, we define the vector field $D_w v$, the *derivative of* v *along* w, by III(8.9). Since $D_w v$ is a vector field, we can similarly define $D_u D_w v = D_u(D_w v)$, provided that all necessary derivatives exist. In general, the derivatives along u and w do not commute: $D_u D_w V \neq D_w D_u v$. One shows from III(8.9) that

$$(D_u D_w v - D_w D_u v - D_{[u,w]} v)^\mu = R^\mu_{\nu\lambda\pi} v^\nu u^\lambda w^\pi, \tag{2.1}$$

where $[u, w]$ is the vector field with components $[u, w]^\mu = u^\pi w^\mu_{,\pi} - w^\pi u^\mu_{,\pi}$ and the $R^\mu_{\nu\lambda\pi}$ are the components in x of the *Riemann tensor field*:

$$R^\mu_{\nu\lambda\pi} = \left\{ {\mu \atop \nu\pi} \right\}_{,\lambda} - \left\{ {\mu \atop \nu\lambda} \right\}_{,\pi} + \left\{ {\rho \atop \nu\pi} \right\}\left\{ {\mu \atop \rho\lambda} \right\} - \left\{ {\rho \atop \nu\lambda} \right\}\left\{ {\mu \atop \rho\pi} \right\}. \tag{2.2}$$

The *Ricci tensor field* is defined by $R_{\nu\pi} = R^\mu_{\nu\mu\pi}$, and the *curvature invariant* by $R = g^{\nu\pi} R_{\nu\pi}$; the *Einstein tensor field* is defined by $G_{\mu\nu} = R_{\mu\nu} - (1/2)Rg_{\mu\nu}$. One can prove that the covariant divergence of the Einstein tensor vanishes (see Appendix D for the idea of covariant divergence). In special relativity, where $R^{\mu}_{\nu\lambda\pi} = 0$, the covariant divergence of the stress-momentum vanishes. If one *assumes* that it always vanishes, even when the Riemann tensor does not vanish, then it is not inconsistent to write $G_{\mu\nu} = 8\pi k T_{\mu\nu}$, where k is a constant. These are the Einstein field equations. More generally, one can write $G_{\mu\nu} + \Lambda g_{\mu\nu} = 8\pi k T_{\mu\nu}$, where Λ is the *cosmological constant*. Note that the vanishing of the covariant divergence of the stress-momentum is a hypothesis: the equation $T^{\nu}_{\mu;\nu} = KR_{,\mu}$, for example, where K is a constant, is equally compatible with the special-relativistic limit [Smalley 83].

Multiplying the Einstein equations by $g^{\mu\nu}$, og $R = -8\pi k g^{\mu\nu}T_{\mu\nu}$, and the equations become $R_{\mu\nu} = 8\pi k(T_{\mu\nu} - (1/2)g_{\mu\nu}g^{\pi\rho}T_{\pi\rho})$. Since $g_{\mu\rho}\{^{\rho}_{\nu\pi}\} = [\nu\pi, \mu]$, eq.(2.2) is equivalent to

$$\begin{aligned} R_{\mu\nu\lambda\pi} &= (1/2)(g_{\mu\pi,\nu\lambda} + g_{\nu\lambda,\mu\pi} - g_{\mu\lambda,\nu\pi} - g_{\nu\pi,\mu\lambda}) \\ &\quad + g^{\sigma\tau}([\mu\pi, \sigma][\nu\lambda, \tau] - [\mu\lambda, \sigma][\nu\pi, \tau]), \end{aligned} \tag{2.3}$$

where $R_{\mu\nu\lambda\pi} = g_{\mu\sigma}R^{\sigma}_{\nu\lambda\pi}$, and the Einstein equations can be written as

$$\begin{aligned} &g^{\mu\lambda}(g_{\mu\pi,\nu\lambda} + g_{\nu\lambda,\mu\pi} - g_{\mu\lambda,\nu\pi} - g_{\nu\pi,\mu\lambda}) + 2g^{\mu\lambda}g^{\sigma\tau}([\mu\pi, \sigma][\nu\lambda, \tau] - [\mu\lambda, \sigma][\nu\pi, \tau]) \\ &\quad = 16\pi k(T_{\nu\pi} - (1/2)g_{\nu\pi}g^{\sigma\rho}T_{\sigma\rho}). \end{aligned} \tag{2.4}$$

To find the postnewtonian limit of (2.4), we assume that $g_{mn} = \delta_{mn} + O_2$, $g_{m0} = O_3$, and $g_{00} = -1 + O_2$, and that the $T_{\mu\nu}$ are given by VII(6.1). We also assume that the gauge conditions VI(2.15) and VI(2.16) hold in the postnewtonian limit (one can prove that this is always possible – the gauge conditions simply restrict the choice of coordinate system).

The Christoffel symbols satisfy $[\mu\nu, \lambda] = L^{-1}O_2$ if none or two of μ, ν, λ are zero, and $[\mu\nu, \lambda] = L^{-1}O_3$ otherwise. To lowest order, the $(0, 0)$ and (m, n) Einstein equations are $g_{00,rr} = -8\pi k\rho c^2 + L^{-2}O_4$, and $g_{mn,rr} = -8\pi k\delta_{mn}\rho c^2 + L^{-2}O_4$, which have the solutions $g_{mn} = \delta_{mn}(1 + 2U) + O_4$ and $g_{00} = -1 + 2U + O_4$. Using these results and the gauge conditions, one proves that the $g_{\mu\nu}$ satisfy VII(6.2), and hence that the Einstein equations have the correct postnewtonian limit.

We have shown that the Einstein equations reduce to the postnewtonian field equations under the appropriate conditions. A harder, because less precise, question is whether they are a *natural* or *plausible* generalization of the postnewtonian equations. If one takes the point of view that gravity is to be described geometrically – specifically, in terms of pseudoriemannian geometry – then one can make a case. The postnewtonian field equations involve the second derivatives of the metric components, and the simplest geometrical quantity that depends on these derivatives is the Riemann tensor. The argument for the Einstein equations may then be developed as above. The actual history of their discovery is summarized in Section 17.7 of [Misner 73].

There is no doubt that the physical world should be partly described in geometrical terms; whether we should strive for a completely geometrical description is still open to debate. Those who favour geometry point to its breadth and elegance; those opposed, to its difficulty and abstractness, and the lack of significant results that cannot be obtained by other means.

3. Vector theories

In VI(3.16) we wrote the metric in a postinertial chart in the explicitly Poincaré covariant form $g_{\mu\nu} = e^{-2\omega}\eta_{\mu\nu} + \alpha n_\mu n_\nu$, where n is a covector field and α and ω are invariants. One can choose ω in many ways, and in particular so that it is a function only of α and the n_μ, as in VI(3.19): $4\omega = \alpha\eta_{\mu\nu}n_\mu n_\nu + O_6$. We derived approximate field equations for the n_μ, or equivalently for the variables $q_m = -n_m/n_0$ and ω or $\beta = \alpha n_0^2$ (eqs.VI(3.27), VI(3.31), etc.). In VII(7.2) and VII(7.3) we expressed these field equations in terms of the stress-momentum of the sources.

It is natural to ask whether one can construct an exact theory of gravity in which the field variables are the n_μ, and the field equations are suitable generalizations of VII(7.2) and VII(7.3). One possibility is to look for a Poincaré covariant theory, in which the components of the metric are the constants $\eta_{\mu\nu}$, and the $g_{\mu\nu}$ are auxiliary quantities – the combinations of the field variables defined by VI(3.16). Another possibility – and the only one that we shall discuss – is to look for a generally covariant theory in which the $g_{\mu\nu}$ play their usual role as components of the spacetime metric.

As a preliminary, we rewrite VI(3.19) in terms of the $g_{\mu\nu}$. We define $n^\mu = g^{\mu\nu}n_\nu$, and the invariant N by

$$N = n^\mu n_\mu = g^{\mu\nu}n_\mu n_\nu. \tag{3.1}$$

Since $\omega = -U + O_4$ and $n_0 = O_1$, $n_m = O_2$, VI(3.19) implies that $4U = \alpha n_0^2 + O_4$, and consequently that $\alpha > 0$. In the postnewtonian approximation, one has $g^{00} = -1 - 2U + O_4, g^{m0} = O_3, g^{mn} = \delta_{mn}(1-2U) + O_4$, and hence $N = \eta_{\mu\nu}n_\mu n_\nu - 2Un_0^2 + O_6$. It follows that $1 - e^{-4\omega} = 4\omega - 8U^2 + O_6 = \alpha\eta_{\mu\nu}n_\mu n_\nu - 8U^2 + O_6 = \alpha N + 2\alpha U n_0^2 - 8U^2 + O_6 = \alpha N + O_6$.

In the exact theory, we assume that ω satisfies the equation that we have just derived, but without an O_6 term:

$$1 - e^{-4\omega} = \alpha N. \tag{3.2}$$

We suppose that there is a chart x' in which the metric has the form VI(3.16): $g'_{\mu\nu} = e^{-2\omega'}\eta_{\mu\nu} + \alpha' n'_\mu n'_\nu$, where the primes denote quantities in x'. In an arbitrary chart x, this becomes

$$g_{\mu\nu} = e^{-2\omega}\bar{g}_{\mu\nu} + \alpha n_\mu n_\nu, \tag{3.3}$$

where the $\bar{g}_{\mu\nu}$ are the components in x of the *flat metric* $\bar{g}$, defined by $\bar{g}_{\mu\nu} = (\partial x'^\pi/\partial x^\mu)(\partial x'^\rho/\partial x^\nu)\eta_{\pi\rho}$.

The components $g^{\mu\nu}$ and $\bar{g}^{\mu\nu}$ are defined to be the solutions of the equations $g^{\mu\nu}g_{\nu\pi} = \delta^\mu_\pi$ and $\bar{g}^{\mu\nu}\bar{g}_{\nu\pi} = \delta^\mu_\pi$, in the usual manner, and we write $\bar{n}^\mu = \bar{g}^{\mu\nu}n_\nu$. To find the equation relating $g^{\mu\nu}$ and $\bar{g}^{\mu\nu}$, we first multiply (3.3) by $\bar{g}^{\pi\mu}n^\nu$, and get $n^\pi = e^{2\omega}(1-\alpha N)\bar{n}^\pi$. We define $\bar{N} = \bar{n}^\mu n_\mu = \bar{g}^{\mu\nu}n_\mu n_\nu$, and (3.2) gives

$$n^\pi = e^{-2\omega}\bar{n}^\pi, \qquad N = e^{-2\omega}\bar{N}. \tag{3.4}$$

Eqs.(3.4) and (3.2) imply that $\alpha\bar{N} = e^{2\omega} - e^{-2\omega} = 4\omega + (8/3)\omega^3 + \ldots$, which is compatible with VI(3.19).

To find the relationship between the $g^{\mu\nu}$ and $\bar{g}^{\mu\nu}$, we first guess that $g^{\mu\nu} = e^{2\omega}\bar{g}^{\mu\nu} + \zeta n^\mu n^\nu$ for some invariant ζ. Eq.(3.3) gives $g^{\mu\nu}g_{\nu\pi} = \delta^\mu_\pi + \alpha e^{2\omega}\bar{n}^\mu n_\pi + \zeta n^\mu n_\pi = \delta^\mu_\pi$, and hence $\alpha e^{4\omega} + \zeta = 0$ from (3.4) and

$$g^{\mu\nu} = e^{2\omega}\bar{g}^{\mu\nu} - \alpha e^{4\omega}n^\mu n^\nu. \tag{3.5}$$

Since $\alpha > 0$ in the postnewtonian limit (see after (3.1)), we can get rid of it by redefining n_μ (i.e. by writing n_μ in place of $\alpha^{1/2}n_\mu$ in (3.3)). Alternatively, we can make (3.3) and (3.5) look neater by taking $\alpha = e^{-2\omega}$. In what follows, we shall be slightly more general and take $\alpha = e^{-a\omega}$, where a is a constant.

The postnewtonian field equations VII(7.2) and VII(7.3) are not obviously covariant. Instead of trying to guess generally covariant generalizations of them, we proceed much as in the Einstein case: we choose some exact field equations on the basis of simplicity, and then compare their postnewtonian limit with VII(7.2) and VII(7.3).

The royal road for inventing field equations is the Lagrangian method. For the rest of this section, we assume that the reader knows the elements of Lagrangian field theory or is prepared to learn it from Appendix G or elsewhere. If he does not and is not, he can still follow the main line of the discussion because the equations are very similar to those of Lagrangian mechanics. The essential difference is that the arguments of the Lagrangian in field theory are functions of the spacetime variables, while in mechanics they are functions of the time only.

We consider a system that is described by functions u_A of the spacetime coordinates (in our case, the u_A might include the n_μ, or alternatively q_m and ω or

β, and variables to describe the matter distribution). We recall that the Lagrangian of the system is an invariant function L of the u_A and their first partial derivatives with respect to the x^μ, which we denote by $u_{A,\mu}$. The Lagrangian may also depend explicitly on the spacetime coordinates, but we need not consider this possibility.

The Lagrangian density $\mathcal{L}$ is defined by $\mathcal{L} = \Delta L$, where $\Delta = |\det(g_{\mu\nu})|^{1/2}$. The Lagrangian density is introduced because its integral over a spacetime region is an invariant, independent of the choice of coordinates. One derives field equations by demanding that this integral be stationary with respect to variation of the u_A – in much the same way that one derives the Lagrangian equations in mechanics by requiring that the integral of the Lagrangian be stationary with respect to variation of the particle variables.

The field equations corresponding to the Lagrangian density $\mathcal{L}$ are

$$(\partial\mathcal{L}/\partial u_A) = (\partial\mathcal{L}/\partial u_{A,\mu})_{,\mu}, \tag{3.6}$$

where $\partial/\partial u_A$ and $\partial/\partial u_{A,\mu}$ denote the partial derivative when the independent variables are the u_A and the $u_{A,\mu}$. As before, the comma denotes the partial derivative when the independent variables are the x^μ (and the u_A are regarded as functions of the x^μ).

Eqs.(3.6) may not be valid for all values of A, but only for a subset $A \in \{1, \ldots, M\}$. For these A, the u_A are called *dynamical variables*; the remaining u_A, say for $A \in \{M+1, \ldots, M+M'\}$, are called *absolute variables.* In our case, the n_μ and the source variables (i.e. the variables that describe the matter distribution) are dynamical variables, and the functions $\bar{g}_{\mu\nu}$ are absolute variables. In the chart x', where the $\bar{g}'_{\mu\nu}$ are the constants $\eta_{\mu\nu}$, it is obvious that they should not be regarded as dynamical variables; it follows that they should not be so regarded in any other chart.

The simplest and most usual assumption in theories of gravity is that the Lagrangian density can be written in the form $\mathcal{L} = \mathcal{L}_G + \mathcal{L}_F$, where $\mathcal{L}_G$ describes the gravitational field and $\mathcal{L}_F$ the sources, and where $\Delta^{-1}\mathcal{L}_G$ and $\Delta^{-1}\mathcal{L}_F$ are invariant functions. More precisely, $\mathcal{L}_F$ is a function of the $g_{\mu\nu}$ and $g_{\mu\nu,\pi}$, and of the source dynamical variables and their first partial derivatives. As for the $\mathcal{L}_G$ term, it is independent of the source dynamical variables: in our case, it can be written as a function of the n_μ and the $\bar{g}_{\mu\nu}$, and of their first partial derivatives $n_{\mu,\pi}$ and $\bar{g}_{\mu\nu,\pi}$.

One usually finds $\mathcal{L}_F$ by writing a known, special relativistic expression in generally covariant form. One has to replace $\eta_{\mu\nu}$ by $g_{\mu\nu}$, and partial derivatives by covariant derivatives – which accounts for the dependence of $\mathcal{L}_F$ on $g_{\mu\nu}$ and $g_{\mu\nu,\pi}$. One can of course then express $g_{\mu\nu}$ and $g_{\mu\nu,\pi}$ in terms of $\bar{g}_{\mu\nu}$, $\bar{g}_{\mu\nu,\pi}$, n_μ, and $n_{\mu,\pi}$ by means of (3.3), and so regard $\mathcal{L}_F$ as a function of these variables. Since the $\bar{g}_{\mu\nu}$ are symmetric, one can take the absolute variables to be the $\bar{g}_{\mu\nu}$ with $\mu \leq \nu$.

Let us refine the notation for partial derivatives. We define ∂ to be the partial derivative when the independent variables are the $n_\mu, n_{\mu,\pi}, \bar{g}_{\mu\nu}$ and $\bar{g}_{\mu\nu,\pi}$ with $\mu \leq \nu$, the source dynamical variables and their first partial derivatives. We define D to be the partial derivative when the independent variables are the $n_\mu, n_{\mu,\pi}, g_{\mu\nu}$ and $g_{\mu\nu,\pi}$ with $\mu \leq \nu$, the source dynamical variables and their first partial derivatives.

We first have to see how the ∂ and D partial derivatives are related to the partial derivatives with respect to the x^μ. Let f be any C^2 function of $u = (u_1, \ldots, u_{M+M'})$ – note that f does *not* depend on the partial derivatives $u_{A,\mu}$, so that $\partial f/\partial u_{A,\mu} = 0$. By the rules for transforming partial derivatives, og

$$\begin{aligned} f_{,\mu} &= (\partial f/\partial u_B) u_{B,\mu}, \\ \partial f_{,\mu}/\partial u_A &= (\partial^2 f/\partial u_A \partial u_B) u_{B,\mu} = (\partial f/\partial u_A)_{,\mu}, \\ \partial f_{,\mu}/\partial u_{A,\pi} &= (\partial f/\partial u_B)(\partial u_{B,\mu}/\partial u_{A,\pi}) = (\partial f/\partial u_A)\delta_{\mu\pi}, \end{aligned} \tag{3.7}$$

where we have written $f_{,\mu}$ instead of $f_{,\mu}(u)$, etc. Eq.(3.7) holds also for the partial derivative D, with the appropriate choice of u_A.

The field equations (3.6) for $u_A = n_\mu$ become

$$\begin{aligned} \partial\mathcal{L}_G/\partial n_\mu - (\partial\mathcal{L}_G/\partial n_{\mu,\pi})_{,\pi} &= -\partial\mathcal{L}_F/\partial n_\mu + (\partial\mathcal{L}_F/\partial n_{\mu,\pi})_{,\pi} \\ &= -\sum_{\sigma\leq\tau}[D\mathcal{L}_F/Dg_{\sigma\tau} - (D\mathcal{L}_F/Dg_{\sigma\tau,\pi})_{,\pi}]\partial g_{\sigma\tau}/\partial n_\mu, \end{aligned} \tag{3.8}$$

where we have used (3.7), the fact that the $g_{\mu\nu}$ are functions of the n_π and the $\bar{g}_{\mu\nu}$, and the previous assumptions about $\mathcal{L}_F$.

In the Lagrangian formalism, the components $T^{\mu\nu}$ of the stress-momentum of the sources are defined by

$$(1/2)\Delta T^{\mu\nu} = \delta\mathcal{L}_F/\delta g_{\mu\nu}, \tag{3.9}$$

where

$$\begin{aligned} \delta\mathcal{L}_F/\delta g_{\mu\nu} &= D\mathcal{L}_F/Dg_{\mu\nu} - (D\mathcal{L}_F/Dg_{\mu\nu,\pi})_{,\pi} \text{ if } \mu = \nu, \\ \delta\mathcal{L}_F/\delta g_{\mu\nu} &= (1/2)\big[D\mathcal{L}_F/Dg_{\mu\nu} - (D\mathcal{L}_F/Dg_{\mu\nu,\pi})_{,\pi}\big] \text{ if } \mu < \nu, \\ \delta\mathcal{L}_F/\delta g_{\mu\nu} &= \delta\mathcal{L}_F/\delta g_{\nu\mu} \text{ if } \mu > \nu. \end{aligned} \tag{3.10}$$

More succinctly, we write $2\zeta\delta\mathcal{L}_F/\delta g_{\mu\nu} = \zeta\Delta T^{\mu\nu} = D\mathcal{L}_F/Dg_{\mu\nu} - (D\mathcal{L}_F/Dg_{\mu\nu,\pi})_{,\pi}$, where $\zeta = 1/2$ if $\mu = \nu$ and $\zeta = 1$ if $\mu < \nu$. The factor 1/2 in the second of eqs.(3.10) allows us to rewrite (3.8) as

$$\partial\mathcal{L}_G/\partial n_\mu - (\partial\mathcal{L}_G/\partial n_{\mu,\pi})_{,\pi} = -(1/2)\Delta T^{\sigma\tau}\partial g_{\sigma\tau}/\partial n_\mu. \tag{3.11}$$

To calculate the $\partial g_{\sigma\tau}/\partial n_\mu$ and Δ terms in (3.11), we use (3.3) with $\alpha = e^{a\omega}$. From (3.2) and (3.4) og $\bar{N} = \bar{n}^\mu n_\mu = \bar{g}^{\mu\nu} n_\mu n_\nu = e^{(2-a)\omega} - e^{-(2+a)\omega}$, and hence

$$\begin{aligned} \partial \bar{N}/\partial n_\mu &= 2\bar{g}^{\mu\nu} n_\mu = 2\bar{n}^\mu = 2e^{2\omega} n^\mu \\ &= \Big[(2-a)e^{(2-a)\omega} + (2+a)e^{-(2+a)\omega}\Big]\partial\omega/\partial n_\mu, \end{aligned} \tag{3.12}$$

$$\partial g_{\sigma\tau}/\partial n_\mu = \Big[-2e^{-2\omega}\bar{g}_{\sigma\tau} + a e^{a\omega} n_\sigma n_\tau\Big]\partial\omega/\partial n_\mu + e^{a\omega}(\delta_{\sigma\mu}n_\tau + \delta_{\tau\mu}n_\sigma). \tag{3.13}$$

One shows from (3.3), (3.2), and (3.4) that in a chart where $\bar{g}_{\mu\nu} = \eta_{\mu\nu}$ one has $\det g_{\mu\nu} = -(e^{-2\omega} + \alpha\bar{N})e^{-6\omega} = -e^{-4\omega}$, and hence

$$\Delta = |\det(g_{\mu\nu})|^{1/2} = e^{-2\omega}. \tag{3.14}$$

Eq.(3.14) does not hold in an arbitrary chart. However, the ratio $\Delta/\bar{\Delta}$ is an invariant, where $\bar{\Delta} = |\det \bar{g}_{\mu\nu}|^{1/2}$. Since $\bar{\Delta} = 1$ in a chart where $\bar{g}_{\mu\nu} = \eta_{\mu\nu}$, it follows that in *any* chart

$$\Delta = \bar{\Delta} e^{-2\omega}. \tag{3.15}$$

Exercise 3.1 Prove (3.15). [Note that the transformation law for the $g_{\mu\nu}$ between two charts is the same as that for the $\bar{g}_{\mu\nu}$.] ■

It is often convenient to write the field equations in terms of the partial derivative D rather than ∂ (i.e. we replace the variables $\bar{g}_{\mu\nu}$ and $\bar{g}_{\mu\nu,\pi}$ by $g_{\mu\nu}$ and $g_{\mu\nu,\pi}$). Og

$$\begin{aligned} \partial\mathcal{L}_G/\partial n_\mu &= D\mathcal{L}_G/Dn_\mu \\ &\quad + \sum_{\sigma\le\tau}\Big[D\mathcal{L}_G/Dg_{\sigma\tau}\partial g_{\sigma\tau}/\partial n_\mu + D\mathcal{L}_G/Dg_{\sigma\tau,\rho}\partial g_{\sigma\tau,\rho}/\partial n_\mu\Big], \end{aligned} \tag{3.16}$$

and a similar expression for $\partial\mathcal{L}_G/\partial n_{\mu,\pi}$. We use (3.7) and (3.9), define $\delta\mathcal{L}_G/\delta g_{\mu\nu}$ in the manner of (3.10), and find that the field equations (3.11) become

$$\begin{aligned} \partial\mathcal{L}_G/\partial n_\mu - (\partial\mathcal{L}_G/\partial n_{\mu,\pi})_{,\pi} &= D\mathcal{L}_G/Dn_\mu - (D\mathcal{L}_G/Dn_{\mu,\pi})_{,\pi} \\ &\quad + \delta\mathcal{L}_G/\delta g_{\sigma\tau}\partial g_{\sigma\tau}/\partial n_\mu \\ &= -(1/2)\Delta T^{\sigma\tau}\partial g_{\sigma\tau}/\partial n_\mu, \end{aligned} \tag{3.17}$$

where $\partial g_{\sigma\tau}/\partial n_\mu$ is given by (3.13).

4. A simple vector theory

We assumed in the last section that the dynamical variables that describe the gravitational field are the components n_μ of a covector field n. We assumed also that the Lagrangian density can be written in the form $\mathcal{L} = \mathcal{L}_F + \mathcal{L}_G$, where $\mathcal{L}_G$ is independent of the source dynamical variables, $\mathcal{L}_F$ depends on the $g_{\mu\nu}$, the source dynamical variables, and their first partial derivatives, and $L_F = \Delta^{-1}\mathcal{L}_F$ and $L_G = \Delta^{-1}\mathcal{L}_G$ are invariants. We did not impose any other conditions on $\mathcal{L}$ (apart from differentiability, which we always assume); the field equations (3.17) are consequently quite general. In what follows, we restrict $\mathcal{L}_F$ by requiring that the stress-momentum (3.9) be that of an ideal fluid, as in earlier chapters. We try to find a simple L_G that gives field equations with the correct postnewtonian limit.

The simplest invariant L_G that gives nontrivial field equations is perhaps

$$L_G = F(\omega)n^{\mu;\nu}n_{\mu;\nu} = F(\omega)g^{\mu\kappa}g^{\nu\lambda}n_{\kappa;\lambda}n_{\mu;\nu}, \tag{4.1}$$

where F is a differentiable, invariant function (as yet undetermined), the semicolon once more denotes the covariant derivative defined by the metric g, and ω is given by (3.1) and (3.2) with $\alpha = e^{a\omega}$. The calculation of the field equations even for this L_G is slightly laborious, although straightforward, and some details are given at the end of the section. One finds that

$$\begin{aligned} &D\mathcal{L}_G/Dn_\mu - (D\mathcal{L}_G/Dn_{\mu,\pi})_{,\pi} \\ &\quad = 2\Delta\big[-(Fn^{\mu;\pi})_{;\pi} + F'n^\mu e^{a\omega}[-a+(4+a)e^{-4\omega}]^{-1}n^{\sigma;\tau}n_{\sigma;\tau}\big], \end{aligned} \tag{4.2}$$

where we have written F for $F(\omega)$, and $F' = dF/d\omega$. (It would be more elegant here to take F to be a function of N, but ω is the natural variable for what follows.) We also have

$$\begin{aligned} 2\Delta^{-1}\delta\mathcal{L}_G/\delta g_{\sigma\tau} = {} & \big[Fg^{\sigma\tau} - 2F'e^{a\omega}(-a+(4+a)e^{-4\omega})^{-1}n^\sigma n^\tau\big]n^{\pi;\rho}n_{\pi;\rho} \\ & + (Fn^\tau n^{\sigma;\rho} + Fn^\sigma n^{\tau;\rho})_{;\rho} + (Fn^\tau n^{\rho;\sigma} + Fn^\sigma n^{\rho;\tau})_{;\rho} \\ & - (Fn^\rho n^{\sigma;\tau} + Fn^\rho n^{\tau;\sigma})_{;\rho} - 2F(n^{\tau;\rho}n^\sigma{}_{;\rho} + n^{\rho;\tau}n_\rho{}^{;\sigma}). \end{aligned} \tag{4.3}$$

The field equation for n^μ is found by substituting (4.2), (4.3), and (3.13) into (3.17).

To derive the field equation for ω, one multiplies the n^μ field equation by n_μ. Since $N = g^{\mu\nu}n_\mu n_\nu = n_\mu n^\mu = e^{-a\omega}(1-e^{-4\omega})$ and $n_\mu(Fn^{\mu;\pi})_{;\pi} = (Fn_\mu n^{\mu;\pi})_{;\pi} - Fn^{\mu;\pi}n_{\mu;\pi} = (1/2)\Delta^{-1}(\Delta Fg^{\pi\tau}N_{,\tau})_{,\pi} - Fn^{\mu;\pi}n_{\mu;\pi}$ from (D14), eqs.(3.17) and (4.2) give

$$\begin{aligned} &\Delta^{-1}(\Delta Fg^{\sigma\tau}N_{,\tau})_{,\sigma} - 2\big\{F + F'Ne^{a\omega}\big[-a+(4+a)e^{-4\omega}\big]^{-1}\big\}n^{\sigma;\tau}n_{\sigma;\tau} \\ &\quad = \big[\Delta^{-1}\delta\mathcal{L}_G/\delta g_{\sigma\tau} + (1/2)T^{\sigma\tau}\big](\partial g_{\sigma\tau}/\partial n_\mu)n_\mu, \end{aligned} \tag{4.4}$$

where from (3.13)

$$(\partial g_{\sigma\tau}/\partial n_\mu)n_\mu$$
$$= \big[-2e^{-2\omega}\bar{g}_{\sigma\tau} + ae^{a\omega}n_\sigma n_\tau\big]2(1-e^{-4\omega})\big[2-a+(2+a)e^{-4\omega}\big]^{-1}$$
$$+ 2e^{a\omega}n_\sigma n_\tau. \tag{4.5}$$

Derivation of eqs.(4.2) and (4.3)

We sketch the derivation of these equations. We recall that the independent variables for the partial derivative ∂ include $\bar{g}_{\sigma\tau}$ and $\bar{g}_{\sigma\tau,\pi}$ with $\sigma \leq \tau$; for the partial derivative D, the corresponding variables are $g_{\sigma\tau}$ and $g_{\sigma\tau,\pi}$ with $\sigma \leq \tau$.

We have already calculated $\partial g_{\sigma\tau}/\partial n_\mu$ in (3.13). the expression for $\partial\omega/\partial n_\mu$ is contained in (3.12):

$$\partial\omega/\partial n_\mu = 2e^{a\omega}n^\mu\big[2-a+(2+a)e^{-4\omega}\big]^{-1}. \tag{4.6}$$

Similarly, differentiating the equations $N = g^{\mu\nu}n_\mu n_\nu = e^{-a\omega}(1-e^{-4\omega})$, og $dN/d\omega = e^{-a\omega}[-a+(4+a)e^{-4\omega}]$ and $DN/Dn_\mu = 2g^{\mu\nu}n_\nu = 2n^\mu = (dN/d\omega)(D\omega/Dn_\mu)$, and hence

$$D\omega/Dn_\mu = 2e^{a\omega}n^\mu\big[-a+(4+a)e^{-4\omega}\big]^{-1}. \tag{4.7}$$

Obviously one has $DN/Dn_{\mu,\pi} = 0$, $D\omega/Dn_{\mu,\pi} = 0$, $D\Delta/Dn_\mu = 0$, and $D\Delta/Dn_{\mu,\pi} = 0$.

The covariant derivative of n is given by $n_{\sigma;\tau} = n_{\sigma,\tau} - \{{}^{\mu}_{\sigma\tau}\}n_\mu$. Since the Christoffel symbols depend only on the $g_{\mu\nu}$ and their first derivatives, og

$$Dn_{\sigma;\tau}/Dn_\mu = -\{{}^{\mu}_{\sigma\tau}\}, \qquad Dn_{\sigma;\tau}/Dn_{\mu,\pi} = \delta_{\sigma\mu}\delta_{\tau\pi}. \tag{4.8}$$

Differentiating (4.1), og

$$\begin{aligned}\Delta^{-1}D\mathcal{L}_G/Dn_\mu &= F'(D\omega/Dn_\mu)n^{\sigma;\tau}n_{\sigma;\tau} - 2Fn^{\sigma;\tau}\{{}^{\mu}_{\sigma\tau}\},\\ \Delta^{-1}D\mathcal{L}_G/Dn_{\mu,\dot{\pi}} &= 2Fn^{\mu;\pi}.\end{aligned} \tag{4.9}$$

We note that $\Delta_{,\pi} = \Delta\{{}^{\mu}_{\mu\pi}\}$ from (D16), and that $(Fn^{\mu;\pi})_{;\pi} = (Fn^{\mu;\pi})_{,\pi} + Fn^{\lambda;\pi}\{{}^{\mu}_{\lambda\pi}\}$ $+ Fn^{\mu;\lambda}\{{}^{\pi}_{\lambda\pi}\}$, and (4.2) follows.

Since $g_{\mu\nu}$ is symmetric, og

$$Dg_{\mu\nu}/Dg_{\sigma\tau} = \zeta(\delta_{\mu\sigma}\delta_{\nu\tau} + \delta_{\mu\tau}\delta_{\nu\sigma}), \tag{4.10}$$

where $\zeta = 1/2$ if $\sigma = \tau$, and $\zeta = 1$ if $\sigma \neq \tau$. Differentiating $g^{\mu\pi}g_{\pi\nu} = \delta^\mu_\nu$, og $(Dg^{\mu\pi}/Dg_{\sigma\tau})g_{\pi\nu} + g^{\mu\pi}Dg_{\pi\nu}/Dg_{\sigma\tau} = 0$ and hence

$$Dg^{\mu\lambda}/Dg_{\sigma\tau} = -g^{\lambda\nu}g^{\mu\pi}Dg_{\pi\nu}/Dg_{\sigma\tau} = -\zeta(g^{\mu\sigma}g^{\lambda\tau} + g^{\mu\tau}g^{\lambda\sigma}), \tag{4.11}$$

Exactly similar equations hold for $\partial\bar{g}_{\mu\nu}/\partial\bar{g}_{\sigma\tau}$ and $\partial\bar{g}^{\mu\nu}/\partial\bar{g}_{\sigma\tau}$.

Eq.(4.11) implies that

$$\begin{aligned} DN/Dg_{\sigma\tau} &= (Dg^{\mu\nu}/Dg_{\sigma\tau})n_\mu n_\nu = -2\zeta n^\sigma n^\tau, \\ D\omega/Dg_{\sigma\tau} &= (d\omega/dN)(DN/Dg_{\sigma\tau}) = 2\zeta e^{a\omega}\Big[a - (4+a)e^{-4\omega}\Big]^{-1} n^\sigma n^\tau. \end{aligned} \tag{4.12}$$

Similarly, og

$$\begin{aligned} \partial\bar{N}/\partial\bar{g}_{\sigma\tau} &= -2\zeta\bar{n}^\sigma\bar{n}^\tau, \\ \partial\omega/\partial\bar{g}_{\sigma\tau} &= (d\omega/d\bar{N})(\partial\bar{N}/\partial\bar{g}_{\sigma\tau}) \\ &= -2\zeta e^{a\omega}\Big[(2-a)e^{2\omega} + (2+a)e^{-2\omega}\Big]^{-1}\bar{n}^\sigma\bar{n}^\tau, \end{aligned} \tag{4.13}$$

where $\bar{n}^\sigma = e^{2\omega}n^\sigma$ from (3.4). Eq.(D8) gives

$$D\Delta/Dg_{\sigma\tau} = \zeta\Delta g^{\sigma\tau}. \tag{4.14}$$

The Christoffel symbols are related by $\{{}^{\mu}_{\pi\rho}\} = g^{\mu\lambda}[\pi\rho,\lambda]$, and (4.11) implies that

$$(D/Dg_{\sigma\tau})\{{}^{\mu}_{\pi\rho}\} = -\zeta(g^{\mu\sigma}\{{}^{\tau}_{\pi\rho}\} + g^{\mu\tau}\{{}^{\sigma}_{\pi\rho}\}). \tag{4.15}$$

Differentiating $n_{\kappa;\rho} = n_{\kappa,\rho} - \{{}^{\lambda}_{\kappa\rho}\}n_\lambda = n_{\kappa,\rho} - [\kappa\rho,\lambda]n^\lambda$, og

$$Dn_{\kappa;\rho}/Dg_{\sigma\tau} = -n_\lambda(D/Dg_{\sigma\tau})\{{}^{\lambda}_{\kappa\rho}\} = \zeta(n^\sigma\{{}^{\tau}_{\kappa\rho}\} + n^\tau\{{}^{\sigma}_{\kappa\rho}\}), \tag{4.16}$$

$$\begin{aligned} Dn_{\kappa;\rho}/Dg_{\sigma\tau,\pi} &= -n^\lambda(D/Dg_{\sigma\tau,\pi})[\kappa\rho,\lambda] \\ &= -(1/2)n^\lambda(D/Dg_{\sigma\tau,\pi})\Big[g_{\kappa\lambda,\rho} + g_{\rho\lambda,\kappa} - g_{\kappa\rho,\lambda}\Big] \\ &= -(1/2)n^\lambda(D/Dg_{\sigma\tau})\Big[g_{\kappa\lambda}\delta_{\pi\rho} + g_{\rho\lambda}\delta_{\pi\kappa} - g_{\kappa\rho}\delta_{\pi\lambda}\Big] \\ &= -(1/2)\zeta n^\lambda\Big[\delta_{\kappa\sigma}\delta_{\lambda\tau}\delta_{\pi\rho} + \delta_{\kappa\tau}\delta_{\lambda\sigma}\delta_{\pi\rho} + \delta_{\rho\sigma}\delta_{\lambda\tau}\delta_{\pi\kappa} \\ &\qquad + \delta_{\rho\tau}\delta_{\lambda\sigma}\delta_{\pi\kappa} - \delta_{\kappa\sigma}\delta_{\rho\tau}\delta_{\pi\lambda} - \delta_{\kappa\tau}\delta_{\rho\sigma}\delta_{\pi\lambda}\Big], \end{aligned} \tag{4.17}$$

where we have used (3.7). We can now calculate $2\zeta\delta\mathcal{L}_G/\delta g_{\sigma\tau} = D\mathcal{L}_G/Dg_{\sigma\tau} - (D\mathcal{L}_G/Dg_{\sigma\tau,\pi})_{,\pi}$ from (4.1) (cf. (3.10)), and so derive (4.3).

Exercise 4.1 Find the field equations and $\delta\mathcal{L}_G/\delta g_{\sigma\tau}$ for the Lagrangian (4.1) with F now assumed to be a constant. [One can proceed as before or, more directly, use (5.10) to calculate $\partial\mathcal{L}_G/\partial n_\mu - (\partial\mathcal{L}_G/\partial n_{\mu,\pi})_{,\pi}$ in (3.17).] ∎

Exercise 4.2 Find the field equations and $\delta\mathcal{L}_G/\delta g_{\sigma\tau}$ for the skewsymmetrized Lagrangian

$$L_G = (1/4)F(\omega)g^{\mu\kappa}g^{\nu\lambda}(n_{\kappa,\lambda} - n_{\lambda,\kappa})(n_{\mu,\nu} - n_{\nu,\mu}). \tag{4.18}$$

[The Lagrangian (4.18) is similar to that of the Maxwell theory; it is an invariant because $n_{\mu,\nu} - n_{\nu,\mu} = n_{\mu;\nu} - n_{\nu;\mu}$. The calculations are easier than the previous ones because no Christoffel symbols appear. You may discover that

$$\begin{aligned} D\mathcal{L}_G/Dn_\mu &- (D\mathcal{L}_G/Dn_{\mu,\pi})_{,\pi} \\ &= \Delta\Big[-(Fn^{\mu;\pi} - Fn^{\pi;\mu})_{;\pi} + (1/2)F'n^\mu e^{a\omega}[-a + (4+a)e^{-4\omega}]^{-1} \\ &\qquad (n^{\kappa;\lambda} - n^{\lambda;\kappa})(n_{\kappa;\lambda} - n_{\lambda;\kappa})\Big], \end{aligned} \tag{4.19}$$

$$\begin{aligned} 2\Delta^{-1}\delta\mathcal{L}_G/\delta g_{\sigma\tau} = (1/4)\Big[&Fg^{\sigma\tau} - 2F'e^{a\omega}(-a + (4+a)e^{-4\omega})^{-1}n^\sigma n^\tau\Big] \\ &(n^{\kappa;\lambda} - n^{\lambda;\kappa})(n_{\kappa;\lambda} - n_{\lambda;\kappa}) \\ &- (1/2)F(n^{\tau;\rho} - n^{\rho;\tau})g^{\kappa\sigma}(n_{\kappa;\rho} - n_{\rho;\kappa}) \\ &- (1/2)F(n^{\sigma;\rho} - n^{\rho;\sigma})g^{\kappa\tau}(n_{\kappa;\rho} - n_{\rho;\kappa}). \end{aligned} \tag{4.20}$$

The field equation for n^μ is found by substituting (4.19), (4.20), and (3.13) into (3.17).] ∎

5. Postnewtonian approximation of vector theories

The Lagrangian L_G, eq.(4.11), contains an arbitrary function F of the variable ω. We now try to choose F so that the theory has the correct postnewtonian limit. As in Chapter VII, we express the postnewtonian equations in terms of the variables ω and $q_m = -n_m/n_0$. We use a chart in which $\bar{g}_{\mu\nu} = \eta_{\mu\nu}$, so that $\Delta = e^{-2\omega}$ from (3.14), and the metric (3.3) (with $\alpha = e^{a\omega}$) becomes

$$g_{\mu\nu} = e^{-2\omega}\eta_{\mu\nu} + e^{a\omega}n_\mu n_\nu. \tag{5.1}$$

The invariant $\bar{N}$ is related to ω through (3.2) and (3.4):

$$\bar{N} = \eta_{\mu\nu}n_\mu n_\nu = e^{-a\omega}(e^{2\omega} - e^{-2\omega}). \tag{5.2}$$

As before, we assume that $n_0 = O_1$ and $n_m = O_2$ in the postnewtonian limit. Eq.(5.2) then implies that $4\omega = -n_0^2 = O_2$, and (5.1) gives

$$g_{mn} = (1-2\omega)\delta_{mn} + O_4, \qquad g_{m0} = O_3, \qquad g_{00} = -1 - 2\omega + O_4, \tag{5.3}$$

$$g^{mn} = (1+2\omega)\delta_{mn} + O_4, \qquad g^{m0} = O_3, \qquad g^{00} = -1 + 2\omega + O_4. \tag{5.4}$$

It follows that the Christoffel symbols $\{{}^{\mu}_{\pi\rho}\}$ and $[\pi\rho,\mu]$ are $L^{-1}O_2$ if none or two of μ,π,ρ are equal to zero, and are $L^{-1}O_3$ otherwise (mnemonic: O or $2, O_2$).

In deriving the postnewtonian equations, we write all n_0 expressions in terms of ω and n_m(or q_m). Eq.(5.2) gives $n_0^2 = n_m n_m - 4\omega + 4a\omega^2 + O_6 = -4\omega + O_4$. Differentiating with respect to x^μ, og

$$\left.\begin{aligned} n_0 n_{0,0} &= -2\omega_{,0} + O_5, \\ n_0^2 n_{0,0}^2 &= 4\omega_{,0}^2 + O_8, \\ \omega n_{0,0}^2 &= -\omega_{,0}^2 + O_8; \end{aligned}\right. \tag{5.5}$$

$$n_0 n_{0,r} = n_m n_{m,r} - 2\omega_{,r} + 4a\omega\omega_{,r} + O_6; \tag{5.6}$$

$$\left.\begin{aligned} n_0^2 n_{0,r} n_{0,r} &= 4\omega_{,r}\omega_{,r} + O_6, \\ \omega n_{0,r} n_{0,r} &= -\omega_{,r}\omega_{,r} + O_6. \end{aligned}\right. \tag{5.7}$$

From (5.6) og $4\omega_{,r}\omega_{,r} = n_0^2 n_{0,r} n_{0,r} - 2n_0 n_{0,r} n_m n_{m,r} - 8a n_0 n_{0,r}\omega\omega_{,r} + O_8$, and hence

$$\omega n_{0,r} n_{0,r} = (q_m q_m + 3a\omega - 1)\omega_{,r}\omega_{,r} + \omega_{,r} n_m n_{m,r} + O_8. \tag{5.8}$$

The covariant derivative of n is given by $n_{\sigma;\tau} = n_{\sigma,\tau} - \{{}^{\mu}_{\sigma\tau}\} n_\mu$. Eqs.(5.3) and (5.4) imply that $\{{}^{0}_{0m}\} = \omega_{,m} + L^{-1}O_4$ (this is the only Christoffel symbol that we need). One shows that

$$\left.\begin{aligned} n_{0;0} &= n_{0,0} + L^{-1}O_4, \qquad n_{m;n} = n_{m,n} + L^{-1}O_4, \\ n_{0;m} &= n_{0,m} - \{{}^{0}_{0m}\} n_0 + L^{-1}O_5 = n_{0,m} - \omega_{,m} n_0 + L^{-1}O_5, \\ n_{m;0} &= n_{m,0} - \omega_{,m} n_0 + L^{-1}O_5; \end{aligned}\right. \tag{5.9}$$

$$\begin{aligned} n^{\mu;\nu} n_{\mu;\nu} &= g^{\mu\kappa} g^{\nu\lambda} n_{\kappa;\lambda} n_{\mu;\nu} \\ &= -n_{0,m} n_{0,m} + n_{0,0}^2 + 2\omega_{,m} n_0 n_{0,m} + n_{m,r} n_{m,r} + L^{-2}O_6. \end{aligned} \tag{5.10}$$

The first term in (5.10) is $L^{-2}O_2$; the rest are $L^{-2}O_4$.

Before we can calculate the postnewtonian approximations of $\Delta^{-1}\delta\mathcal{L}_G/\delta g_{\sigma\tau}$ and the field equations, we must make some assumption about the function F.

A constant F does not work (see Exercise 5.1). The next simplest possibility is $F(\omega) = \lambda\omega$, where λ is a constant. We neglect the O_6 terms in (4.3), and find that

$$
\begin{aligned}
2\lambda^{-1}\Delta^{-1}\delta\mathcal{L}_G/\delta g_{00} & \\
&= (\omega + (1/2)n_0^2)n_{0,r}n_{0,r} + 2(\omega n_0 n_{0,r})_{,r} - 2\omega n_{0,r}n_{0,r} \\
&\quad + L^{-2}O_6 \\
&= -\omega_{,r}\omega_{,r} - 4\omega\omega_{,rr} + L^{-2}O_6, \\
2\lambda^{-1}\Delta^{-1}\delta\mathcal{L}_G/\delta g_{m0} &= -4q_m\omega\omega_{,rr} - [4\omega^2(q_{r,m} + q_{m,r})]_{,r} + 2\omega_{,m}\omega_{,0} + L^{-2}O_7, \\
2\lambda^{-1}\Delta^{-1}\delta\mathcal{L}_G/\delta g_{mn} &= -\omega\delta_{mn}n_{0,r}n_{0,r} + 2\omega n_{0,m}n_{0,n} + L^{-2}O_6 \\
&= \delta_{mn}\omega_{,r}\omega_{,r} - 2\omega_{,m}\omega_{,n} + L^{-2}O_6,
\end{aligned}
\tag{5.11}
$$

where we have used (5.5)–(5.8) to eliminate the n_0 terms. From (4.5) og

$$(\partial g_{\sigma\tau}/\partial n_\mu)n_\mu = (-4\omega + 8\omega^2 - 4a\omega^2)\eta_{\sigma\tau} + (2 + 4a\omega)n_\sigma n_\tau + O_6. \tag{5.12}$$

To find the postnewtonian approximation of the ω field equation (4.4), we use (5.11) and (5.12), and recall that $T^{00} = \rho c^2 O_0$, $T^{m0} = \rho c^2 O_1$, and $T^{mn} = \rho c^2 O_2$. Since $n_m = -n_0 q_m$, og $n_{m,r} = -n_{0,r}q_m - n_0 q_{m,r}$, $n_m n_{m,r} = n_0 n_{0,r} q_m q_m + n_0^2 q_m q_{m,r}$, $n_{m,r}n_{m,r} = n_{0,r}n_{0,r}q_m q_m + n_0^2 q_{m,r}q_{m,r} + 2n_0 n_{0,r} q_m q_{m,r}$. Again using (5.5)–(5.8) to eliminate the n_0 terms, og

$$
\begin{aligned}
&4\lambda\Big[\omega\eta_{\mu\nu}\omega_{,\mu\nu} - (2 + 2a)\omega^2\omega_{,mm} + 4\omega^2 q_{m,r}q_{m,r} - (3 + (3/2)a)\omega\omega_{,m}\omega_{,m}\Big] \\
&= 8\lambda\omega^2\omega_{,mm} - 2\omega T^{00}(1 + 2\omega + a\omega + 2q_m q_m) \\
&\quad + 8\omega T^{m0}q_m - 2\omega T^{mm} + \lambda L^{-2}O_8.
\end{aligned}
\tag{5.13}
$$

Dividing by ω, we find that to lowest order $4\lambda\omega_{,mm} = -2T^{00} + \lambda L^{-2}O_4$. Choosing $\lambda = -(8\pi k)^{-1}$, and noting that $T^{00} = \rho c^2(1 + O_2)$, og $\omega_{,mm} = 4\pi k\rho c^2(1 + O_2)$, and hence $\omega = -U + O_4$ from (C33). We could of course have deduced this condition directly by requiring that (5.3) be compatible with the postinertial metric, but it is comforting that it follows from the field equations.

Writing $\lambda = -(8\pi k)^{-1}$ in (5.13) and eliminating the $\omega_{,mm}$ terms, og

$$
\begin{aligned}
&\eta_{\mu\nu}\omega_{,\mu\nu} + 4\omega q_{m,r}q_{m,r} - (3 + (3/2)a)\omega_{,m}\omega_{,m} \\
&\qquad = 4\pi k\Big[T^{00}(1 + 6\omega + 3a\omega + 2q_m q_m) - 4T^{m0}q_m + T^{mm}\Big] + L^{-2}O_6.
\end{aligned}
\tag{5.14}
$$

If we choose $a = -2$, eq.(5.14) reduces to the postinertial field equation VII(7.4).

If one puts $F(\omega) = \lambda\omega$ in (4.2) and (4.3) and uses (5.10), og

$$\begin{aligned} D\mathcal{L}_G/Dn_m - (D\mathcal{L}_G/Dn_{m,\pi})_{,\pi} &= -2\Delta\Big[(\lambda\omega n_{m,r})_{,r} + (1/4)n_m n_{0,r} n_{0,r}\Big] \\ &\quad + \lambda L^{-2}O_6, \\ \Delta^{-1}\delta\mathcal{L}_G/\delta g_{\sigma\tau} &= O_4. \end{aligned} \tag{5.15}$$

Eqs.(3.12) and (3.13) give $\partial g_{\sigma\tau}/\partial n_m = (-\eta_{\sigma\tau}n_m + \delta_{\sigma m}n_\tau + \delta_{\tau m}n_\sigma)(1 + O_2)$, and the field equation (3.17) becomes

$$\begin{aligned} &\lambda\Big[(\omega n_{m,r})_{,r} + (1/4)n_m n_{0,r} n_{0,r}\Big] \\ &\quad = (1/4)T^{\sigma\tau}(-\eta_{\sigma\tau}n_m + \delta_{\sigma m}n_\tau + \delta_{\tau m}n_\sigma) + \lambda L^{-2}O_6. \end{aligned} \tag{5.16}$$

We write $\omega = -(1/4)n_0^2 + O_4$, use (5.6), and show that the right-hand side is $-(1/4)\lambda\Big[n_0(n_0 n_m)_{,rr} + 2\omega_{,rr}n_m + L^{-2}O_6\Big]$. We set $\lambda = -(8\pi k)^{-1}$ and $\omega_{,rr} = 4\pi k\rho c^2(1 + O_2)$ and find that

$$(n_0 n_m)_{,rr} = 16\pi k T^{m0} + L^{-2}O_5, \tag{5.17}$$

which agrees with VII(7.2). We have therefore proved that if $F(\omega) = -\omega(8\pi k)^{-1}$ and $a = -2$, then the Lagrangian (4.1) gives field equations with the standard, postnewtonian limit. Since $n_0 n_m = 4\omega q_m + O_5$, eq.(5.17) is equivalent to $(\omega q_m)_{,rr} = 4\pi k T^{m0} + L^{-2}O_5$.

Exercise 5.1 Calculate the postnewtonian approximation of the field equations found in Exercise 4.1. ∎

Exercise 5.2 Calculate the postnewtonian approximation of the field equations found in Exercise 4.2. ∎

A more general choice of F

Instead of taking $F(\omega) = \lambda\omega$, let us suppose that $F(\omega) = \lambda(\omega + b\omega^2)$, where b is a constant. The b dependent terms in $\Delta^{-1}\delta\mathcal{L}_G/\delta g_{\sigma\tau}$ are O_4. The b dependent terms that are added to the left-hand side of (5.13) are $4\lambda\Big[b\omega^2\omega_{,mm} + (1/2)b\omega\omega_{,m}\omega_{,m}\Big]$, and the ω field equation (5.14) becomes

$$\begin{aligned} &\eta_{\mu\nu}\omega_{,\mu\nu} + 4\omega q_{m,r}q_{m,r} - (1/2)(6 + 3a - b)\omega_{,m}\omega_{,m} \\ &\quad = 4\pi k\Big\{T^{00}\Big[1 + (6 + 3a - b)\omega + 2q_m q_m\Big] - 4T^{m0}q_m + 2T^{mm}\Big\} \\ &\quad\quad + L^{-2}O_6. \end{aligned} \tag{5.18}$$

There is no change in the n_m field equation (5.17).

Eqs.(5.17) and (5.18) remain valid for any F that can be expanded as a power series in ω (the terms of degree greater than two contribute nothing to the postnewtonian field equations).

When solving (5.18), we can replace ω by $-U$ in all but the first term. We know from the discussion of perihelion shifts in Chapter IV, Section 5, that the parameter $6+3a-b$ must be fairly small, $|6+3a-b| \ll 1$. On the right-hand side of (5.18), we can interpret $k[1+(6+3a-b)\omega]$ as a potential-dependent gravitational constant. According to many cosmological theories, the average value of ω (or U) is a function of time; the present strength of the gravitational force would therefore differ from that in the early universe. A value $|6+3a-b| < 1$ would not however produce observable cosmological effects ([Misner 73], §40.8).

The multiplicity of exact theories

We can now see more clearly that the postnewtonian theory cannot be uniquely and plausibly generalized to give an exact theory of gravity. We have shown that, if gravity is to be described in terms of a covector field or in terms of the Riemann tensor, then there are simple, exact theories that have the correct postnewtonian limit. These theories are however quite different from one another, and they are different again from many other exact theories that have been proposed on the basis of reasonable mathematical or physical assumptions. There are insufficient logical grounds for preferring any one of the theories, and choices on the basis of aesthetics are too subjective to be given much credence. A reliable, exact theory must await new experimental results.

6. Energy and momentum

Energy and momentum are easily defined in most special-relativistic field theories. One introduces a stress-momentum – which is a symmetric tensor field T obtained by a standard prescription from the Lagrangian of the system. The ordinary divergence of T vanishes for an isolated system: $T^{\mu\nu}{}_{,\nu} = 0$, where the comma denotes the partial derivative with respect to the coordinates of an inertial chart. The components of the 4 momentum of the system at the instant $x^0 = ct$ are defined to be $P^\mu(x^0) = c^{-1} \int T^{\mu 0}(x^0, \boldsymbol{x}) d^3x$, where the integral is over all $\boldsymbol{x}$. If the integral exists, the vanishing of the divergence of T implies that the P^μ are constants, independent of t. The energy of the system is cP^0.

For field theories in curved spacetime, where special relativity is not valid, it may be impossible to define conserved energy and momentum. If the theory has a Lagrangian, one can still introduce a stress-momentum T, but the divergence equation $T^{\mu\nu}{}_{,\nu} = 0$ does not hold. Under certain conditions, which will be stated in a moment, the *covariant* divergence of T does vanish: $T^{\mu\nu}{}_{;\nu} = 0$. This does not imply that $\int T^{\mu 0}(x^0, \boldsymbol{x})\, d^3x$ are constants, but one can often define conserved energy and momentum by adding suitable terms to the $T^{\mu\nu}$, as we did in Chapter VII, Section 5. We write $\Theta^{\mu\nu} = T^{\mu\nu} + t^{\mu\nu}$, and try to find $t^{\mu\nu}$ so that the ordinary divergence of Θ vanishes. It can be done in the Einstein theory if the spacetime is asymptotically flat, for example (i.e. if spacetime differs very little from flat spacetime at large spatial distances from the origin).

The proof that $T^{\mu\nu}{}_{;\nu} = 0$ is contained in Appendix G. It is assumed that the Lagrangian density can be written in the form $\mathcal{L} = \mathcal{L}_F + \mathcal{L}_G$, where $\mathcal{L}_F = \Delta L_F$ and $\mathcal{L}_G = \Delta L_G$, and L_F and L_G are invariants. As usual, $\Delta = |\det(g_{\mu\nu})|^{1/2}$ and g is the spacetime metric. In the manner of Section 3, we write the dynamical variables as $u_1, \ldots, u_M$ and the absolute variables as $u_{M+1}, \ldots, u_{M+M'}$. We assume that $\mathcal{L}_F$ is a function of the dynamical variables $u_1, \ldots, u_P$ and their first partial derivatives, and of the $g_{\mu\nu}$ and $g_{\mu\nu,\pi}$ with $\mu \leq \nu$, and of no other variables. (Note that $P \leq M$; in applying the theorem we shall assume that $u_1, \ldots, u_P$ are the source dynamical variables.) The $g_{\mu\nu}$ and $\mathcal{L}_G$ may be functions of $u_{P+1}, \ldots, u_{M+M'}$ and their partial derivatives, but do not depend on $u_1, \ldots, u_P$. We define T by $\Delta T^{\mu\nu} = 2\delta\mathcal{L}_F/\delta g_{\mu\nu}$, eq.(3.9), and prove that $T^{\mu\nu}{}_{;\nu} = 0$, eq.(G31).

Many field theories can be written in terms of a Lagrangian with the above properties; the Einstein and vector theories of gravity are two examples. One can however easily construct theories for which $T^{\mu\nu}{}_{;\nu} \neq 0$.

We are going to show how energy and momentum are defined in the vector theories of Sections 3 and 4. It is straightforward because they can be regarded as flat-spacetime theories – the metric $\bar{g}$ has components $\eta_{\mu\nu}$ in a suitable chart. We define a stress-momentum Θ directly in terms of the Lagrangian and the metric $\bar{g}$, and show that the ordinary divergence of Θ vanishes in a chart where $\bar{g}_{\mu\nu} = \eta_{\mu\nu}$.

We change the notation in the obvious manner, writing the Lagrangian density as $\mathcal{L} = \bar{\Delta}\bar{L} = \bar{\mathcal{L}}_F + \bar{\mathcal{L}}_G$, where $\bar{\Delta} = |\det(\bar{g}_{\mu\nu})|^{1/2}$. (Our notation does not distinguish between $\mathcal{L}$ as a function of $g_{\mu\nu}$, etc., and $\mathcal{L}$ as a function of $\bar{g}_{\mu\nu}$, etc.) Since $\mathcal{L} = \Delta L$, og $\bar{L} = \bar{\Delta}^{-1}\Delta L = e^{-2\omega}L$ from (3.15). Note that $\mathcal{L}$ is the same as before, but $\bar{L}$ is different from L – it is $\mathcal{L}$ that determines the field equations. The dynamical variables $u_1, \ldots u_M$ are the variables that describe the sources together with the n_μ(or ω and q_m). The only absolute variables are the $\bar{g}_{\mu\nu}$ for $\mu \leq \nu$, corresponding to $u_{M+1}, \ldots, u_{M+10}$, say.

The stress-momentum Θ is defined by

$$\bar{\Delta}\Theta^{\mu\nu} = 2\delta\mathcal{L}/\delta\bar{g}_{\mu\nu}, \tag{6.1}$$

It is proved in Appendix G (see (G27) and the discussion that follows) that $\Theta^{\mu\nu}{}_{|\nu} = 0$, where the vertical bar denotes the covariant derivative corresponding to the metric $\bar{g}$. In a chart where $\bar{g}_{\mu\nu} = \eta_{\mu\nu}$, this covariant derivative reduces to a partial derivative, and one has $\Theta^{\mu\nu}{}_{,\nu} = 0$, eq.(G28). One can therefore define energy and momentum exactly as in a special relativistic theory.

There is a simple relationship between the variational derivatives with respect to g and $\bar{g}$ that allows us to calculate Θ in terms of the stress-momentum T, where $\Delta T^{\mu\nu} = 2\delta\mathcal{L}_F/\delta g_{\mu\nu}$. The variational derivatives are given by equations analogous to (3.10) (cf. after (4.17)):

$$\begin{aligned} 2\zeta\delta\mathcal{L}/\delta g_{\sigma\tau} &= D\mathcal{L}/Dg_{\sigma\tau} - (D\mathcal{L}/Dg_{\sigma\tau,\pi})_{,\pi}, \\ 2\zeta\delta\mathcal{L}/\delta\bar{g}_{\sigma\tau} &= \partial\mathcal{L}/\partial\bar{g}_{\sigma\tau} - (\partial\mathcal{L}/\partial\bar{g}_{\sigma\tau,\pi})_{,\pi}, \end{aligned} \tag{6.2}$$

where $\zeta = 1/2$ if $\sigma = \tau$, and $\zeta = 1$ if $\sigma < \tau$. The rules for transforming partial derivatives imply that

$$\begin{aligned} \partial\mathcal{L}/\partial\bar{g}_{\sigma\tau} &= \sum_{\mu\leq\nu}\Big[D\mathcal{L}/Dg_{\mu\nu}\partial g_{\mu\nu}/\partial\bar{g}_{\sigma\tau} - D\mathcal{L}/Dg_{\mu\nu,\rho}\partial g_{\mu\nu,\rho}/\partial\bar{g}_{\sigma\tau}\Big], \\ \partial\mathcal{L}/\partial\bar{g}_{\sigma\tau,\pi} &= \sum_{\mu\leq\nu} D\mathcal{L}/Dg_{\mu\nu,\rho}\partial g_{\mu\nu,\rho}/\partial\bar{g}_{\sigma\tau,\pi}. \end{aligned} \tag{6.3}$$

From (3.7) og $\partial g_{\mu\nu,\rho}/\partial\bar{g}_{\sigma\tau} = (\partial g_{\mu\nu}/\partial\bar{g}_{\sigma\tau})_{,\rho}$, $\partial g_{\mu\nu,\rho}/\partial\bar{g}_{\sigma\tau,\pi} = \partial g_{\mu\nu}/\partial\bar{g}_{\sigma\tau}\delta_{\rho\pi}$, and hence

$$\partial\mathcal{L}/\partial\bar{g}_{\sigma\tau} - (\partial\mathcal{L}/\partial\bar{g}_{\sigma\tau,\pi})_{,\pi} = \sum_{\mu\leq\nu}\Big[D\mathcal{L}/Dg_{\mu\nu} - (D\mathcal{L}/Dg_{\mu\nu,\pi})_{,\pi}\Big]\partial g_{\mu\nu}/\partial\bar{g}_{\sigma\tau}.$$

From (6.2), this is equivalent to

$$2\zeta\delta\mathcal{L}/\delta\bar{g}_{\sigma\tau} = \delta\mathcal{L}/\delta g_{\mu\nu}\partial g_{\mu\nu}/\partial\bar{g}_{\sigma\tau}. \tag{6.4}$$

Since $\mathcal{L} = \mathcal{L}_F + \mathcal{L}_G$, eqs.(6.1) and (3.9) give

$$\zeta\bar{\Delta}\Theta^{\mu\nu} = \Big[(1/2)\Delta T^{\mu\nu} + \delta\mathcal{L}_G/\delta g_{\mu\nu}\Big]\partial g_{\mu\nu}/\partial\bar{g}_{\sigma\tau}. \tag{6.5}$$

Eq.(6.5) is quite general: it requires only that the $g_{\mu\nu}$ depend algebraically on the $\bar{g}_{\sigma\tau}$, so that (6.3) and (3.7) hold.

We now specialize to the case when (3.3) holds with $\alpha = e^{a\omega}$. Differentiating (3.3), og $\partial g_{\mu\nu}/\partial\bar{g}_{\sigma\tau} = e^{-2\omega}\partial\bar{g}_{\mu\nu}/\partial\bar{g}_{\sigma\tau} + (-2e^{-2\omega}\bar{g}_{\mu\nu} + ae^{a\omega}n_\mu n_\nu)\partial\omega/\partial\bar{g}_{\sigma\tau}$, and (4.13) and the $\bar{g}$ analog of (4.10) imply that

$$\begin{aligned} \partial g_{\mu\nu}/\partial\bar{g}_{\sigma\tau} &= \zeta e^{-2\omega}(\delta_{\mu\sigma}\delta_{\nu\tau} + \delta_{\mu\tau}\delta_{\nu\sigma}) \\ &\quad + 2\zeta e^{a\omega}(2e^{-2\omega}\bar{g}_{\mu\nu} - ae^{a\omega}n_\mu n_\nu)\Big[(2-a)e^{2\omega} + (2+a)e^{-2\omega}\Big]^{-1}\bar{n}^\sigma\bar{n}^\tau. \end{aligned} \tag{6.6}$$

From (6.5), using $\Delta = \bar{\Delta} e^{-2\omega}$ from (3.15) and noting that $T^{\mu\nu}$ and $\delta\mathcal{L}_G/\delta g_{\mu\nu}$ are symmetric, og

$$\Theta^{\sigma\tau} = e^{-4\omega}T^{\sigma\tau} + 2e^{-4\omega}\Delta^{-1}\delta\mathcal{L}_G/\delta g_{\sigma\tau} + e^{-2\omega}\Big[T^{\mu\nu} + 2\Delta^{-1}\delta\mathcal{L}_G/\delta g_{\mu\nu}\Big]$$
$$e^{a\omega}\bar{n}^\sigma\bar{n}^\tau(2e^{-2\omega}\bar{g}_{\mu\nu} - ae^{a\omega}n_\mu n_\nu)\Big[(2-a)e^{2\omega} + (2+a)e^{-2\omega}\Big]^{-1}. \tag{6.7}$$

If $a = -2$, this becomes

$$\Theta^{\sigma\tau} = e^{-4\omega}T^{\sigma\tau} + (1/2)e^{-2\omega}T^{\mu\nu}g_{\mu\nu}n^\sigma n^\tau + 2e^{-4\omega}\Delta^{-1}\delta\mathcal{L}_G/\delta g_{\sigma\tau}$$
$$+ e^{-2\omega}\Delta^{-1}(\delta\mathcal{L}_G/\delta g_{\mu\nu})g_{\mu\nu}n^\sigma n^\tau. \tag{6.8}$$

We easily find the postnewtonian limit of (6.8) by using (5.11). We recall that $\omega = -U + O_4$, $\lambda = -(8\pi k)^{-1}$, and $\omega_{,mm} = 4\pi k T^{00} + L^{-2}O_4$, and find that (6.8) agrees with VII(5.6) provided that terms of order O_5 are negligible.

Exercise 6.1 Calculate the O_5 terms in Θ^{m0} and compare with those in VII(5.6). [You should find that they agree apart from a spatial divergence:

$$\Theta^{m0} = T^{m0} - (4\pi k)^{-1}U_{,m}U_{,0} + (2\pi k)^{-1}\Big[U^2(q_{r,m} - q_{m,r})\Big]_{,r} + k^{-1}L^{-2}O_7. \tag{6.9}$$

The proof uses the field equation (5.17), which implies that $(\omega q_m)_{,rr} = 4\pi k T^{m0} + L^{-2}O_5$, and hence that $4\omega q_m T^{00} - 4\omega T^{m0} = -(\pi k)^{-1}(\omega^2 q_{m,r})_{,r} + k^{-1}L^{-2}O_7$.] ■

Because the field equations of the vector theory have the correct postnewtonian limit, we know that the $\Theta^{\mu\nu}$ of VII(5.6) are conserved. However, we have not proved that the $\Theta^{\mu\nu}$ are unique (there might be other quantities $\Psi^{\mu\nu}$ that are conserved). It is therefore not redundant to show that the conserved quantities (6.7) have VII(5.6) as their postnewtonian limit.

7. Gravitational radiation

We know that accelerated, electrically charged bodies lose energy in the form of electromagnetic radiation. By analogy, one might expect that accelerated, massive bodies would lose energy in the form of gravitational radiation, and this is indeed predicted by most exact theories of gravity. The rate at which gravitational energy is radiated is almost always extremely small, and its effect on the motion of massive bodies is unobservable over short time periods. It has not yet been directly detected on Earth, although some changes in the orbital parameters of the binary pulsar PSR 1913+16 are ascribed to it.

The postnewtonian theory is insufficiently precise for the calculation of gravitational radiation. However, it is useful – and indeed almost essential – as an *adjunct* to calculations with exact theories. We shall describe the role that it plays, and outline some of the difficulties of the calculations.

In addition to its use in the calculation of gravitational radiation, the postnewtonian theory enables us to disentangle the observable effects of the radiation. We saw in Chapter VIII, eq.(1.20), that the expression for the energy of a system in the postnewtonian theory can be interpreted in terms of the Newtonian, kinetic and potential energies of the gravitating bodies. Suppose that we have an exact theory of gravity in which the energy reduces to the postnewtonian expression in the postnewtonian limit. When this approximation is valid, we can interpret changes in the energy (as calculated from the theory) in terms of changes in the Newtonian energies of the gravitating bodies, and hence in terms of changes in their orbits. The postnewtonian theory therefore acts as an intermediary between the exact theory and observation.

The changes in energy in the exact theory may be calculated, for example, from the flux of gravitational energy across a large sphere that surrounds the sources. The changes in the orbits are thereby related to the radiation of gravitational energy.

The calculation of gravitational radiation is easy to outline but hard to accomplish. One first constructs a stress-momentum Θ that has the correct postnewtonian limit, and whose ordinary divergence vanishes in a suitable coordinate system. One assumes that the sources of the gravitational field are spatially bounded: they are contained in a ball of radius R_1 centred at the spatial origin. By integrating the equation $\Theta^{\mu\nu}_{,\nu} = 0$ over a ball of radius $R > R_1$ and applying the divergence theorem, og

$$(d/dt)\int_{r\le R}\Theta^{00}(x^0,\boldsymbol{x})\,d^3x = c\int_{r=R}\Theta^{0m}(x^0,\boldsymbol{x})x^m R\sin\theta\,d\theta\,d\phi, \tag{7.1}$$

where $r = |\boldsymbol{x}|, \theta$, and ϕ are spherical polar coordinates. The rate of energy loss by gravitational radiation is defined to be the limit of the right-hand side of (7.1) as $R \to \infty$.

The problem of calculating gravitational radiation has now been reduced to that of finding $\Theta^{0m}(x^0,\boldsymbol{x})$ for large values of $r = |\boldsymbol{x}|$. Since Θ^{0m} is a known function of the gravitational field variables, this is equivalent to finding solutions of the gravitational field equations valid for large r.

The field equations of exact theories are always complicated, and to solve them easily one must make fairly drastic simplifications. The crudest procedure is to linearize them. If one neglects terms of second or higher degree in the field variables, and products of the field variables and the $T^{\mu\nu}$, the field equations usually reduce to inhomogeneous wave equations or something equally simple. This primitive procedure may be quite adequate for the calculation of the gravitational waves produced by a rotating, rigid body, for example.

If the theory is one of the many for which $T^{\mu\nu}{}_{;\nu} = 0$, the assumptions of the last paragraph imply that the Christoffel-symbol terms in $T^{\mu\nu}{}_{;\nu}$ are negligible, and therefore that $T^{\mu\nu}{}_{,\nu} = 0$ in a suitable chart. For a rigid body, this causes no difficulty. However, for an ideal fluid we recall that the equation $T^{\mu\nu}{}_{;\nu} = 0$ implies that the paths of the fluid particles are geodesics of the metric. The equation $T^{\mu\nu}{}_{,\nu} = 0$ therefore implies that the paths are geodesics of the flat spacetime metric – which means that the velocity of each fluid particle is constant. We conclude that it is inconsistent to use the linearized field equations to calculate the radiation from fluid bodies moving under their mutual gravitational attraction.

The traditional way to calculate radiation from gravitating bodies is to keep the linear terms in the field equations while replacing the nonlinear terms by their postnewtonian approximations. There are some technical problems because the support of the postnewtonian terms is spatially unbounded, but the method seems plausible. However, it is difficult to show that the results are correct – that the small terms which are neglected do not contribute significantly to the even smaller radiation terms.

To obtain reliable results, a more serious mathematical campaign must be mounted. In an inner region, for $r < R_2$ say, one expands the solution of the exact field equations in powers of c^{-1}. The first terms of the series are the usual Newtonian and postnewtonian approximations, but the series is not convergent in general: it is an asymptotic series for the solution in the inner region. In an outer region, for $r > R_1$ say, one uses a weak-field approximation, and expresses the solution of the field equation as a multipole expansion. One chooses $R_1 < R_2$ and matches the inner and outer solutions in the region $R_1 < r < R_2$ where both expansions are valid. This allows one to express the field variables for large values of r in terms of integrals over the sources.

The above procedure has been carried through for the Einstein theory, where it confirms and extends the results of the naive postnewtonian method [Blanchet 88]. A more elegant way of calculating gravitational radiation is to abandon the postnewtonian theory and look for an approximation method that is valid everywhere, not just in a neighbourhood of the sources. An example of such a method – again for the Einstein theory – is described in [Schäfer 85]. In all cases the calculations are substantial, and we shall not describe them further. For more information on gravitational radiation, mainly in the context of the Einstein theory, the articles by Damour and Thorne in [Hawking 87] should be consulted. There has so far been little rigorous work on radiation in other exact theories of gravity.

The rate of loss of gravitational energy, as calculated from the Einstein theory, is given in first approximation by the celebrated *quadrupole formula*:

$$DE_{\text{rad}} = -(G/45c^4)\langle D^3Q^{mn}D^3Q^{mn}\rangle, \tag{7.2}$$

where D is the time derivative and Q^{mn} is the (m,n) component of the quadrupole moment of the radiating body as defined by VIII(1.18) (many authors denote the quadrupole moment by $I^{mn} = Q^{mn}/3$). The pointed brackets in (7.2) indicate a time average over several periods of rotation or oscillation. Eq.(7.2) agrees with the observed change in the orbital parameters of the binary pulsar PSR 1913+16. So far, this is the only observation of gravitational radiation.

We remark that the motion of the binary pulsar is described quite well by the postnewtonian theory, even though it consists of two neutron stars for which the postnewtonian theory cannot be expected to apply! A fairly general scheme – the postkeplerian formalism – has been devised which allows one to compare the motion of such binary systems with the predictions of exact theories of gravity [Will 81], [Damour 86, 88].

Notes

[1]To make this plausible, suppose that y is a dimensionless function of the spacetime variables that satisfies the homogeneous wave equation in flat spacetime. One then has $(\partial_0 y)^2 = \partial_m y \partial_m y$, and $\partial_m y = L^{-1}O_r$ implies that $\partial_0 y = L^{-1}O_r$ – which contradicts the postnewtonian assumption that $\partial_0 y = L^{-1}O_{r+1}$. The result also holds in an approximately flat spacetime.

Afterword

You have been guided through the pleasant shallows of Newtonian theory and gravitostatics and down the treacherous rapids of postnewtonian calculation. You float at last on the broad estuary of exact theory that leads, if one can find the channel, to the open sea. Or so it is rumoured.

I shall leave you here and go ashore. You must decide whether to continue on or turn back. Before committing yourself, and if the last chapters were not too intimidating, you should take a few day-trips on the [Schutz 85], the [Wald 84], or one of the other boats. As for becoming a professional gravitator, or even a Ph.D. student, caution is advisable. Experiments in gravity are difficult and long; it is noble work, but not for the faint-hearted. Theoretical gravity is less forbidding and perhaps less noble. You can see the theorists out there, bobbing around in large groups and shouting at one another. Join them if you wish. However, to make real discoveries you will have to sail off alone or with a few companions. Expect derision.

Exact theories are beguiling but inconclusive, like unfinished poems:

> Beyond Styx a season of pomegranates
> Blood upon lips and dark promises ...

We wait at the cave mouth for Delphic hints, and scan the sky for black holes and dark matter. Searching for what we expect to see, we fabricate universes:

> The huge decisions printed out by feet
> Inventing where they tread,
> The random windows conjuring a street. – [Larkin 74]

This is the virtue of theories – right or wrong they direct attention to what may be observed; they illuminate the possible. We must treat them seriously but ironically, like brilliant but eccentric friends. The proper strategy is to develop a few of them in some depth. Not just one, because that will limit our insight, and lead us to confuse quirks of a particular theory with necessary features of the universe. Not too many, or we shall lack the strength for deep penetration.

But I am anticipating the pleasures ashore. Goodbye!

A Kindergarten Relativity

If you have never studied relativity, this appendix should enable you to read at least Chapters I–IV. It may help those also who have studied the subject and been confused.

We recall from Chapter I that in Newtonian physics there are preferred, rectangular coordinate systems called *Galilean charts* and preferred time coordinates, and that a Galilean chart together with a preferred time coordinate is a *Galilean frame.* Coordinate transformations that relate Galilean frames are called *Galilean transformations.* A simple example of such a transformation is

$$x'^1 = x^1 - Vt, \qquad x'^2 = x^2, \qquad x'^3 = x^3, \qquad t' = t, \tag{A1}$$

where V is a constant, $\boldsymbol{x} = (x^1, x^2, x^3)$ and $\boldsymbol{x}' = (x'^1, x'^2, x'^3)$ are rectangular cartesian coordinates, and t and t' are time coordinates. Other examples of Galilean transformations are shifts of the spatial or temporal origin, spatial rotations, and reflections of the space or time axes.

In an informal way, we speak of $\boldsymbol{x} = (x^1, x^2, x^3)$ as a *spatial point* of a Galilean chart or frame. If the x^m are constant, independent of the time coordinate, we say that the spatial point is *fixed* in the Galilean frame.

Special relativity resembles Newtonian physics in having preferred charts in which the descriptions of physical systems are particularly simple. They are called *inertial charts.* They assign three cartesian coordinates and a time coordinate, and are therefore analogous to Galilean frames rather than Galilean charts. However, they are not the same as Galilean frames: they are related by *Poincaré transformations*, which are in general different from Galilean transformations. Some Galilean transformations are in fact also Poincaré transformations – for example, shifts of the spatial or temporal origin, spatial rotations, and reflections of the space or time axes. But the Galilean transformation (A1) is *not* a Poincaré transformation. We are going to derive the simple Poincaré transformation analogous to (A1).

To make the argument less abstract, we imagine that our laboratory is fixed in the chart K. We can think of spatial points of K as corresponding to material particles at rest in K, each attached to a small clock. We may speak of a clock fixed in K as a *K clock*, and a measuring rod fixed in K as a *K measuring rod.*

We assume that K is an inertial chart. This means that it satisfies three conditions. First, the clocks fixed in K can be set so that their readings always correspond to the time coordinate of K. Second, the Euclidean distance $|\boldsymbol{x} - \boldsymbol{y}| = [(x^1 - y^1)^2 + (x^2 - y^2)^2 + (x^3 - y^3)^2]^{1/2}$ between any two spatial points $\boldsymbol{x}$ and $\boldsymbol{y}$ of K is equal to the distance between the corresponding particles as measured by a K measuring rod. Finally, the speed of light in vacuum as measured in K has always the same value c.

We define a chart $\overline{K}$ whose spatial points all have the same constant velocity $\boldsymbol{V} = (V, 0, 0)$ as measured in K (note that we allow negative values of V – we are not defining $V = |\boldsymbol{V}|$. Just as in the case of K, we can think of the spatial points of $\overline{K}$ as corresponding to material particles with small clocks attached, and we use similar terminology – $\overline{K}$ clocks and measuring rods, etc. The $\overline{K}$ clocks and measuring rods all have the same constant velocity $\boldsymbol{V}$ as measured in K. The time coordinate $\bar{t}$ of $\overline{K}$ is defined so that $\bar{t} = 0$ when $t = 0$. That is, each $\overline{K}$ clock is set to read zero at the instant when it is close to a K clock that reads zero.

We shall not discuss the experimental evidence for special relativity (the life-times of elementary particles, the Michelson-Morley experiment, etc.), which you can read in the relativity textbooks. One finds that the results of the experiments can be derived from two assumptions. –

(i) As measured in the chart K, a clock fixed in $\overline{K}$ goes slower by a factor $\gamma = (1 - |\boldsymbol{V}|^2 c^{-2})^{-1/2}$ than exactly similar clocks fixed in K.

(ii) If a measuring rod is fixed in the chart $\overline{K}$, the component of its length in the direction of $\boldsymbol{V}$ is smaller by a factor γ when measured in the chart K than when measured in $\overline{K}$. The component of its length perpendicular to $\boldsymbol{V}$ is the same in K and $\overline{K}$.

To interpret (i), we note that if a clock goes slower, it makes fewer ticks and gives a smaller measure to a time interval. Similarly, for (ii), if a measuring rod is shrunk, it gives a larger measure to a distance. Assumption (ii) is a statement of the *FitzGerald contraction.*

The clocks of $\overline{K}$ have been set to agree with those of K when $t = 0$. It follows from (i) that, at the instant t in K, the clocks of $\overline{K}$ read $\bar{t} = t/\gamma$ (each clock in $\overline{K}$ is compared with the K clock that is near it at the instant t as measured by the K clock).

We can define the spatial coordinates of $\overline{K}$ so that the length, measured in

$\overline{K}$, of a measuring rod at rest in $\overline{K}$ with endpoints $\bar{\boldsymbol{x}}$ and $\bar{\boldsymbol{y}}$ is $|\bar{\boldsymbol{x}} - \bar{\boldsymbol{y}}|$. To do this, let us suppose that a spatial point of $\overline{K}$ (or a particle fixed in $\overline{K}$) has spatial coordinates x_0^1, x_0^2, x_0^3 in K at the instant $t = 0$. Its spatial coordinates in $\overline{K}$ at $t = 0$ are defined to be $\bar{x}^1 = \gamma x_0^1$, $\bar{x}^2 = x_0^2$, $\bar{x}^3 = x_0^3$. The spatial point is fixed in $\overline{K}$, so its spatial coordinates $\bar{x}^m$ are constant in time; it has a constant velocity $(V, 0, 0)$ in K, so its spatial coordinates in K at the instant t are $x^1 = x_0^1 + Vt$, $x^2 = x_0^2$, $x^3 = x_0^3$. We therefore have

$$\bar{x}^1 = \gamma(x^1 - Vt), \qquad \bar{x}^2 = x^2, \qquad \bar{x}^3 = x^3, \qquad \bar{t} = t/\gamma. \tag{A2}$$

If $\bar{\boldsymbol{x}}$ and $\bar{\boldsymbol{y}}$ are the spatial coordinates in $\overline{K}$ of the endpoints of a measuring rod fixed in $\overline{K}$, and $\boldsymbol{x}$ and $\boldsymbol{y}$ are their spatial coordinates in K at the instant t, then (A2) gives

$$\begin{aligned} |\bar{\boldsymbol{x}} - \bar{\boldsymbol{y}}| &= \left[(\bar{x}^1 - \bar{y}^1)^2 + (\bar{x}^2 - \bar{y}^2)^2 + (\bar{x}^3 - \bar{y}^3)^2\right]^{1/2} \\ &= \left[\gamma^2(x^1 - y^1)^2 + (x^2 - y^2)^2 + (x^3 - y^3)^2\right]^{1/2}. \end{aligned} \tag{A3}$$

The distance in K between the spatial points with coordinates $\boldsymbol{x}$ and $\boldsymbol{y}$ is $|\boldsymbol{x} - \boldsymbol{y}| = [(x^1-y^1)^2+(x^2-y^2)^2+(x^3-y^3)^2]^{1/2}$, and this is the length in K of the rod fixed in $\overline{K}$ (i.e. one measures the distance in K between the ends of the rod at the same instant t in K). If we multiply the $(x^1 - y^1)^2$ term by a factor γ^2, we get the right-hand side of (A3). We therefore interpret $|\bar{\boldsymbol{x}} - \bar{\boldsymbol{y}}|$ as the length of the rod as measured in $\overline{K}$, and (A3) then gives a precise, mathematical expresssion of Assumption (ii).

We have assumed that the speed of light in vacuum always has the same value c when measured in the inertial chart K. This means that if a light signal is emitted from the spatial point with coordinates $\boldsymbol{x}$ at the instant t, and arrives at the spatial point with coordinates $\boldsymbol{y}$ at the instant s, then $|\boldsymbol{x} - \boldsymbol{y}|^2 = c^2(t - s)^2$. One shows from (A2) that an equation of this form does not hold in $\overline{K}$, and the speed of light is therefore not constant as measured in $\overline{K}$. However, we can make it constant by a simple resynchronization of clocks. We define a new chart K' with the same spatial coordinates as $\overline{K}$, but a different time coordinate:

$$x'^1 = \bar{x}^1, \qquad x'^2 = \bar{x}^2, \qquad x'^3 = \bar{x}^3, \qquad t' = \bar{t} - \bar{x}^1 V/c^2. \tag{A4}$$

This means that the readings of a clock fixed in $\overline{K}$ at the spatial point $\overline{K}$ are changed by the (constant) amount $-\bar{x}^1 V/c^2$.

Combining (A2) and (A4), og

$$\begin{gathered} x'^1 = \gamma(x^1 - Vt), \qquad x'^2 = x^2, \qquad x'^3 = x^3, \\ t' = \gamma(t - x^1 V/c^2), \end{gathered} \tag{A5}$$

and hence

$$|\boldsymbol{x}'|^2 - c^2t'^2 = |\boldsymbol{x}|^2 - c^2t^2. \tag{A6}$$

If in addition the coordinates $\boldsymbol{y}, s$ and $\boldsymbol{y}', s'$ are related by equations exactly similar to (A5) (so that $y'^1 = \gamma(y^1 - Vs)$, etc.), og

$$|\boldsymbol{x}' - \boldsymbol{y}'|^2 - c^2(t' - s')^2 = |\bar{\boldsymbol{x}} - \bar{\boldsymbol{y}}|^2 - c^2(t - s)^2. \tag{A7}$$

In the special case when a light signal is emitted from the spatial point with coordinates $\boldsymbol{x}$ at the instant t in K, and arrives at the spatial point with coordinates $\boldsymbol{y}$ at the instant s in K, the right-hand side of (A7) vanishes (since the speed of light is c in K). It follows that the speed of light in K' always has the same constant value c.

We have established that K' satisfies the three criteria for an inertial chart. The proof was based on assumptions (i) and (ii), and the requirement that K be an inertial chart. We conclude that the coordinate transformation (A5) is a Poincaré transformation (it transforms the coordinates of an inertial chart into those of another inertial chart). We define a Poincaré transformation to be a *Lorentz transformation* if it leaves the space and time origins unchanged (i.e. $\boldsymbol{x} = 0$ and $t = 0$ imply that $\boldsymbol{x}' = 0$ and $t' = 0$). Since (A5) has this property, it is a Lorentz transformation.

Solving (A5) we find that

$$\begin{gathered} x^1 = \gamma(x'^1 + Vt'), \qquad x^2 = x'^2, \qquad x^3 = x'^3, \\ t = \gamma(t' + x'^1 V/c^2), \end{gathered} \tag{A8}$$

which has exactly the same form as (A5) except for the exchange of primed and unprimed variables and the replacement of V by $-V$.

We can now go backwards through our previous argument. We define a chart $\tilde{K}$ by $\boldsymbol{x} = \tilde{\boldsymbol{x}}$, $t = \tilde{t} + \tilde{x}^1 V/c^2$, and find that

$$\tilde{x}^1 = \gamma(x'^1 + Vt'), \qquad \tilde{x}^2 = x'^2, \qquad \tilde{x}^3 = x'^3, \qquad \tilde{t} = t'/\gamma. \tag{A9}$$

This allows us to deduce (i) and (ii), but with the charts K and $\overline{K}$ replaced by K' and $\tilde{K}$, respectively. We have therefore shown that the charts K and K' are both inertial, and that the relationships between moving and fixed clocks and measuring rods are the same in both of them, as expressed by the appropriate versions of (i) and (ii).

The constancy of the speed of light in an inertial chart is intimately related to the possibility of synchronizing clocks. We must first define what this means. Let us suppose that a light signal is emitted at the instant t_1 from the clock C_1. It is reflected without delay at the clock C_2 at the instant t_2, and returns to C_1 at the instant t_1'. The signal travels in vacuum throughout; t_1 and t_1' are measured by C_1, and t_2 by C_2. We say that C_2 is *synchronized* with C_1 at the instant $(1/2)(t_1 + t_1')$ if $t_2 = (1/2)(t_1 + t_1')$. The last equation is equivalent to $t_2 - t_1 = t_1' - t_2$. Note that the definition of synchronization does not make any reference to a chart; in particular, it is not required that the clocks be fixed in an inertial chart.

If C_2 is synchronized with C_1 at the instant t, it is *not* true in general that it is synchronized at other instants, nor that C_1 is synchronized with C_2. However, if C_1 and C_2 are fixed in the same inertial chart, and both are set to read the time coordinate of the chart, then each is synchronized with the other at every instant. To show this, one has only to note that the distance between the clocks is constant, and so is the speed of light signals from one to the other. It follows that the time required for a light signal to travel from the first clock to the second is the same as that required for it to travel from the second to the first – which means that $t_2 - t_1 = t_1' - t_2$, in the notation of the last paragraph. One should also note that if the clocks C_1, C_2, and C_3 are all fixed in the same inertial chart, and C_2 is synchronized with C_1, and C_3 is synchronized with C_2, then C_3 is synchronized with C_1. We may say that the constancy of the speed of light in the inertial chart allows us to *consistently* synchronize the clocks that are fixed in the chart.

The derivation of the Lorentz transformation (A5) is in the spirit of Lorentz,[1] but differs from those found in most elementary relativity texts. After reviewing experiments, they usually assert something that appears irrational: that the speed of light is the same in every inertial chart. This implies that, no matter how fast you run after a light signal, it continues to move away from you at exactly the same speed – which is very hard to swallow! They then derive the Lorentz transformations (perhaps making some assumption about linearity), and show that the experimental results are indeed explained.

It is in fact all quite logical, but it exacts a terrible price. It erodes a student's trust in physical intuition, and encourages a blind reliance on abstract thinking which is quite unnecessary. One can derive the Lorentz transformations, as we have shown, from hypotheses that do not violate common sense, and then *prove* the constancy of the speed of light. The proof reveals the hidden assumptions about synchronization, etc., and makes the result unparadoxical.

When one has understood the meaning of the constancy of the speed of light, it is enlightening and amusing to turn the argument around and make this a hypothesis of the theory. It is sadistic and pedagogically insane to inflict such sophistication on naive students.

If you have studied relativity before, you can ignore the following exercises. If the subject is new to you, they may be helpful.

Exercise A1 Two small objects with a wire stretched between them move on the x^1 axis of an inertial chart K in such a manner that the difference between their x^1 coordinates is constant. The objects start from rest in K. After a period of acceleration, they move at constant speed $V > 0$ in K. Is the wire then stretched more, or less, or the same amount as it was initially? Explain your answer! ■

Exercise A2 A thin rod of proper length 1.1m moves parallel to the x^1 axis of an inertial chart K at a constant speed $0.8c$. A thin, circular ring, whose normal is parallel to the x^2 axis of K, moves in the x^2 direction with a constant speed W. The proper diameter of the ring is 1m.

(a) Show that, with a suitable choice of initial conditions, it is possible for the rod to pass through the ring.

(b) Describe what happens in an inertial chart K' where the rod is stationary. Does the rod still go through the ring? If so, how? If not, why not? Draw a diagram showing the positions of the objects at an instant t' in K'.

[Recall that a *proper* quantity is one measured in a chart where the object is (instantaneously) at rest.] ■

Exercise A3 An opaque cylinder has a circular cross-section of diameter d when at rest in an inertial chart. It moves parallel to a plane wall at a constant velocity $\mathbf{V}$ in an inertial chart K. The axis of the cylinder is parallel to the wall and perpendicular to $\mathbf{V}$. The cylinder is illuminated by parallel light that shines perpendicular to the wall as measured in K. Find the width of the cylinder's shadow in K. ■

Exercise A4 Generalize Exercise A3 by considering any opaque object instead of a cylinder. Show that the shadow in K has the same shape as when the object is stationary in K but rotated with respect to its original orientation. Find the axis and angle of the rotation. ■

Notes

[1] The transformation was first derived by W. Voigt in 1887, and then by P. Larmor in 1900. It became widely known through the work of Lorentz, and was named after him by H. Poincaré [Lorentz 37].

B Poincaré Transformations

Introduction

In Appendix A, we derived a simple form of the Lorentz transformation. We are now going to find the most general coordinate transformation between inertial charts – the general Poincaré transformation. It would be possible, although not simple, to extend the previous argument to this case, but we shall take another path. In Chapter II, eq.II(8.10), we introduced the interval function – which is a function of the coordinates of *two* spacetime points. Eq.(A7) shows that the interval function is invariant under the simple Lorentz transformation (A5). We now turn the argument around and make the invariance of the interval function the *defining* criterion for a Poincaré transformation.

We first recall that a spacetime chart assigns a unique, ordered quadruple of real numbers to each spacetime point in some region, with the restriction that no quadruple is assigned to more than one point. We are going to consider charts for which the spacetime region (which is called the *domain* of the chart) is the whole of spacetime, and for which each ordered quadruple is assigned to some point. To be more mathematical, we consider a chart $K : \mathcal{T} \to \mathbb{R}^4$, $K(p) = x$, where K is a bijection (a one-to-one and onto map), $\mathcal{T}$ is the set of spacetime points, $\mathbb{R}^4$ is the set of ordered quadruples of real numbers, and $x = (x^0, x^1, x^2, x^3)$ are the coordinates (in K) of the spacetime point p. The x^m are the spatial coordinates of p, and x^0/c is the time coordinate.[1] If $K' : \mathcal{T} \to \mathbb{R}^4$, $K'(p) = x'$, is another chart, then K' is a bijection, and the coordinates x and x' are related by $x' = K' \circ K^{-1}(x) = B(x)$, where the map $B : \mathbb{R}^4 \to \mathbb{R}^4$ is a bijection. We call B a *coordinate transformation (from K to K')*, and say that x and x' are the coordinates of the same spacetime point in K and K', respectively.

Unless otherwise stated, lower-case Latin indices have the range $\{1,\ 2,\ 3\}$ and lower-case Greek indices have the range $\{0,\ 1,\ 2,\ 3\}$. Both kinds of index obey the summation convention: when an index is repeated in a term, sum over its range.

The general Poincaré transformation

If x and y are the coordinates of any spacetime points in the chart K, the *interval function* I is defined by

$$I(x,y) = \eta_{\mu\nu}(x^\mu - y^\mu)(x^\nu - y^\nu). \tag{B1}$$

where $\eta_{mn} = \delta_{mn}, \eta_{\mu 0} = \eta_{0\mu} = -\delta_{\mu 0}$. Eq.(B1) is the same as II(8.10), except that we are now using the summation convention.

A map $P : I\!\!R^4 \to I\!\!R^4$ is defined to be a *Poincaré transformation* if $I(P(x), P(y)) = I(x,y)$ for all $x, y \in I\!\!R^4$. This definition, together with a differentiability condition, allows us to determine explicitly the form of Poincaré transformations and to prove that they are bijections – and hence coordinate transformations as defined above.

For the following theorem, we recall that a function is C^N if all its Nth order partial derivatives exist and are continuous, and that if a function f is C^N with $N \geq 2$, its partial derivatives satisfy $\partial_\mu \partial_\nu f = \partial_\nu \partial_\mu f$. For any map $F : I\!\!R^4 \to I\!\!R^4$, $F(x) = x' = (x'^0, x'^1, x'^2, x'^3)$, we define functions $F^\mu : I\!\!R^4 \to I\!\!R$ by $F^\mu(x) = x'^\mu$ for all x, and write $F = (F^0, F^1, F^2, F^3)$. We define F to be C^N if all the F^μ are C^N.

Theorem If P is a C^2 Poincaré transformation, then there are $\lambda, P^\mu_\nu \in I\!\!R$ such that $\eta_{\mu\nu} P^\mu_\pi P^\nu_\rho = \eta_{\pi\rho}$ and $(P(x))^\mu = P^\mu_\nu x^\nu + \lambda^\mu$ for all $x \in I\!\!R^4$.

Proof Define functions $P^\mu : I\!\!R^4 \to I\!\!R$ such that $P = (P^0, P^1, P^2, P^3)$. Since $I(P(x), P(y)) = I(x,y)$, eq.(B1) gives

$$\eta_{\mu\nu}[P^\mu(x) - P^\mu(y)][P^\nu(x) - P^\nu(y)] = \eta_{\mu\nu}(x^\mu - y^\mu)(x^\nu - y^\nu). \tag{B2}$$

Differentiate (B2) with respect to x^π, and use $\eta_{\mu\nu} = \eta_{\nu\mu}$:

$$\eta_{\mu\nu}\partial_\pi P^\mu(x)[P^\nu(x) - P^\nu(y)] = \eta_{\pi\nu}(x^\nu - y^\nu). \tag{B3}$$

Differentiate (B3) with respect to y^σ and then with respect to x^ρ:

$$\eta_{\mu\nu}\partial_\pi P^\mu(x)\partial_\sigma P^\nu(y) = \eta_{\pi\sigma}, \tag{B4}$$

$$\eta_{\mu\nu}\partial_\rho\partial_\pi P^\mu(x)\partial_\sigma P^\nu(y) = 0. \tag{B5}$$

We can solve (B5) for the $\partial_\rho\partial_\pi P^\mu(x)$. We shall write the determinant of the matrix M with elements $M_{\mu\nu}$ as $\det M$ or $\det(M_{\mu\nu})$, as convenient. If we regard (B4) as a matrix equation, and apply the usual rule for the determinant of a product of matrices, we find

$$\det(\partial_\pi P^\mu(x)) \det(\eta_{\mu\nu}\partial_\sigma P^\nu(y)) = \det(\eta_{\pi\sigma}) = -1.$$

It follows that $\det(\eta_{\mu\nu}\partial_\sigma P^\nu(y)) \neq 0$, and the matrix with elements $\eta_{\mu\nu}\partial_\sigma P^\nu(y)$ has an inverse. Multiplying (B5) by this inverse, og $\partial_\rho\partial_\pi P^\mu(x) = 0$ for all x. The general, C^2 solution of this system of equations is

$$P^\mu(x) = P^\mu_\nu x^\nu + \lambda^\mu, \tag{B6}$$

where $P^\mu_\nu, \lambda^\mu \in I\!R$. It follows that $\partial_\pi P^\mu(x) = P^\mu_\pi$, and from (B4) og

$$\eta_{\mu\nu} P^\mu_\pi P^\nu_\rho = \eta_{\pi\rho}, \tag{B7}$$

which completes the proof. ■

We have shown that any Poincaré transformation can be written in the form (B6), with the P^μ_ν satisfying (B7), provided that it is C^2 (i.e. the second-order partial derivatives of the P^μ exist and are continuous). The C^2 assumption is unnecessarily strong: it is only necessary to assume that the transformation is continuous [Nanda 76]. From now on, we assume that every Poincaré transformation can be written as in (B6) and (B7).

Lorentz matrices

A real, 4×4 matrix L whose elements L^μ_ν satisfy $\eta_{\mu\nu} L^\mu_\pi L^\nu_\rho = \eta_{\pi\rho}$ is called a *Lorentz matrix*. Eq.(B6) shows that every Poincaré transformation determines a unique Lorentz matrix. The *transpose* of L is the matrix L^T whose (μ,ν) element is $L^{T\mu}_\nu = L^\nu_\mu$. We can then write the defining equation for a Lorentz matrix in the equivalent forms

$$L^T \eta L = \eta, \qquad \eta_{\mu\nu} L^\mu_\pi L^\nu_\rho = \eta_{\pi\rho}, \tag{B8}$$

where $\eta = (\eta_{\mu\nu})$.

Since $\det\eta = -1$ and $\det L^T = \det L$, eq.(B8) implies that $(\det L)^2 = 1$, $\det L = \pm 1$, and the inverse L^{-1} of L exists. Multiplying (B8) by η, og $\eta L^T \eta L = \eta^2 = I_4$, where I_4 is the 4×4 unit matrix, and the inverse of L is therefore $L^{-1} = \eta L^T \eta$. Multiplying by L from the left[2] gives $L\eta L^T \eta = I_4$, and hence

$$L\eta L^T = \eta, \qquad \eta^{\mu\nu} L^\pi_\mu L^\rho_\nu = \eta^{\pi\rho}, \tag{B9}$$

where we have written $\eta^{\mu\nu} = \eta_{\mu\nu}$. Eqs.(B9) are equivalent to (B8).

The Poincaré group

A Poincaré transformation P is a *Lorentz transformation* if it leaves the spacetime origin invariant: that is if $P(0) = 0$. A Poincaré transformation is a *translation* if its Lorentz matrix is the identity matrix. It is easy to prove that Lorentz transformations and translations are bijections from $I\!R^4$ to $I\!R^4$. Since a Poincaré transformation can always be written as the composition of a Lorentz transformation and a translation, any Poincaré transformation is a bijection from $\mathbb{R}^4$ to $\mathbb{R}^4$. The Poincaré transformations are therefore coordinate transformations, as defined earlier.

The translations obviously form a group. One readily shows that the Lorentz matrices and the Lorentz transformations each form a group, and that these groups are isomorphic.[3] In what follows, we may call either of them the *Lorentz group*. The Poincaré transformations form a group, the *Poincaré group*, of which the Lorentz group and the group of translations are subgroups.

For any Lorentz matrix L, one has $|\det L| = 1$ and $|L^0_0| \geq 1$ (note that $(L^0_0)^2 = 1 + L^0_m L^0_m$, from (B9)). A Lorentz matrix is *proper* (resp. *improper*) if $\det L = 1$(resp. $\det L = -1$).[4] A Lorentz matrix is *orthochronous* (resp. *anorthochronous*) if $L^0_0 \geq 1$ (resp. $L^0_0 \leq -1$). The proper Lorentz matrices form a group, as do the orthochronous Lorentz matrices. A Lorentz matrix that is both proper and orthochronous is said to be *restricted*. The restricted Lorentz matrices form a group – the *restricted* Lorentz group – which is the intersection of the proper and orthochronous groups.

The terminology for Lorentz transformations is exactly similar. The proper, orthochronous, and restricted groups of Lorentz transformations are defined in the obvious way in terms of the Lorentz matrices of the transformations. We shall normally use the same name for a group of transformations and the isomorphic group of matrices.

Exercise B1 Define $x \in \mathbb{R}^4$ to be *positive* if $x^0 > 0$ and *timelike* if $\eta_{\mu\nu}x^\mu x^\nu < 0$. Prove that the Lorentz transformation $L : \mathbb{R}^4 \to \mathbb{R}^4$ is orthochronous iff it maps a positive, timelike x onto a positive, timelike y. ∎

Exercise B2 Use the result of the last exercise to prove that the orthochronous Lorentz transformations form a group. ∎

We define *space reflection* to be the map $(x^0, x^1, x^2, x^3) \mapsto (x^0, -x^1, -x^2, -x^3)$, and *time inversion* to be the map $(x^0, x^1, x^2, x^3) \mapsto (-x^0, x^1, x^2, x^3)$, for all $x \in \mathbb{R}^4$. Space reflection is an improper Lorentz transformation, and any improper Lorentz transformation can be written as the composition of space reflection and a proper Lorentz transformation. Similarly, time inversion is an anorthochronous Lorentz transformation, and any anorthochronous Lorentz transformation can be written as the composition of time inversion and an orthochronous Lorentz transformation.

Exercise B3 Prove the assertions in the last paragraph. ∎

The reader is by now thoroughly tired of scholastic classifications. For our purposes, the essential point is that the most general Poincaré transformation can be written as the composition of a restricted Lorentz transformation L (with $\det L = 1$ and $L_0^0 \geq 1$), and possibly space reflection, time reversal, and a translation. The last three can often be dealt with very easily. In the remainder of this appendix, we consider restricted Lorentz transformations almost exclusively.

Velocity of one chart in another

We have introduced the Lorentz matrices in a formal, mathematical way, without much explanation of their physical significance. You may recall that the elements of a 3×3 rotation matrix can be expressed in a rather complicated fashion in terms of the Euler angles. We are going to show that the L_0^μ and L_μ^0 elements of a Lorentz matrix have a much simpler interpretation.

A spatial point of the chart K is determined by the values Y^m of its spatial coordinates in K. The Y^m are constants, independent of the time coordinate x^0 of K, and we can identify a spatial point of K with a line parallel to the x^0 axis. Similarly, a spatial point of the chart K' is determined by the values Z'^m of its spatial coordinates in K', where the Z'^m are constants independent of x'^0, and we can identify the spatial point with a line parallel to the x'^0 axis.

We assume that the charts K and K' with coordinates x and x', respectively, are related by a restricted Lorentz transformation L, so that $x' = L(x)$, or $x'^\mu = L_\nu^\mu x^\nu$. The inverse transformation is $x = L^{-1}(x')$ where $L^{-1} = \eta L^T \eta$, or equivalently $x^\mu = \eta^{\mu\lambda} L_\lambda^\pi \eta_{\pi\nu} x'^\nu$. The spacetime point with coordinates x^0, Y^m in K has coordinates x'^0, Y'^m in K', where

$$x'^0 = L_0^0 x^0 + L_p^0 Y^p, \qquad Y'^m = L_0^m x^0 + L_p^m Y^p. \tag{B10}$$

If x^0 changes by δx^0 and the Y^p stay constant, the changes in x'^0 and Y'^m are $\delta x'^0 = L_0^0 \delta x^0$ and $\delta Y'^m = L_0^m \delta x^0$. We write $V'^m = c\delta Y'^m/\delta x'^0 = cL_0^m/L_0^0$, and define $\boldsymbol{V}'$ to be the 3 vector with components V'^m in K'. We say that $\boldsymbol{V}'$ is the 3 velocity in K' of a spatial point of K(or of a spatial point fixed in K). Since $\boldsymbol{V}'$ is independent of the Y^p, it is the same for all spatial points of K, and we say that it is the 3 velocity of K in K'.

Similarly, since $\eta_{mn} = \delta_{mn}$, $\eta_{\mu 0} = \eta_{0\mu} = -\delta_{\mu 0}$, and $\eta^{\mu\nu} = \eta_{\mu\nu}$, the spacetime point with coordinates x'^0, Z'^m in K' has coordinates x^0, Z^m in K, where

$$x^0 = L_0^0 x'^0 - L_0^p Z'^p, \qquad Z^m = -L_m^0 x'^0 + L_m^p Z'^p \tag{B11}$$

If x'^0 changes by $\delta x'^0$ and the Z'^p stay constant, the changes in x^0 and Z^m are $\delta x^0 = L_0^0 \delta x'^0$ and $\delta Z^m = -L_m^0 \delta x'^0$. We write $V^m = c\delta Z^m/\delta x^0 = -cL_m^0/L_0^0$, and define $\boldsymbol{V}$ to be the 3 vector with components V^m in K. We say that $\boldsymbol{V}$ is the 3 velocity in K of a spatial point of K' (or of a spatial point fixed in K'), or that it is the 3 velocity of K' in K.

In summary, we have shown that

$$V'^m = cL^m_0/L^0_0, \qquad V^m = -cL^0_m/L^0_0, \tag{B12}$$

where $\boldsymbol{V}'$ is the 3 velocity of K in K', and $\boldsymbol{V}$ is the 3 velocity of K' in K. The *speed* of K in K' is $V' = |\boldsymbol{V}'|$, and from (B8) og $(V')^2 = c^2(L^{02}_0 - 1)/L^{02}_0$. The *speed* of K' in K is $V = |\boldsymbol{V}|$, and from (B9) og $V = V'$. Since L is a restricted Lorentz matrix, and hence orthochronous, one has $L^0_0 \geq 0$ and therefore $L^0_0 = \gamma$, where $\gamma = (1 - |\boldsymbol{V}|^2/c^2)^{-1/2}$. Substituting in (B12), og

$$L^m_0 = \gamma V'^m/c, \qquad L^0_m = -\gamma V^m/c, \qquad L^0_0 = \gamma. \tag{B13}$$

The velocities $\boldsymbol{V}$ and $\boldsymbol{V}'$ therefore determine and are determined by the $(m, 0)$, $(0, m)$, and $(0, 0)$ elements of the restricted Lorentz matrix L.

Spatial rotations and boosts

A restricted Lorentz transformation is still a fairly complicated object. The best way to understand it is to first consider two special cases – spatial rotations and boosts – which are intuitively easy to grasp. We then prove that an arbitrary, restricted Lorentz transformation can be written as the product of a spatial rotation and a boost.

A spatial rotation is defined to be a restricted Lorentz transformation with $L^0_0 = 1$ (and hence $L^0_m = L^m_0 = 0$, from (B8) and (B9)). The spatial rotations map (1, 0, 0, 0) onto itself: they leave the time coordinate (or time axis) unchanged, and rotate the spatial axes. They form a subgroup of the restricted Lorentz group that is isomorphic to the group of 3 dimensional rotations.

A *boost* is a restricted Lorentz transformation with a symmetric Lorentz matrix: $L^\mu_\nu = L^\nu_\mu$. This algebraic characterization does not completely define boosts (see Exercise B4), nor does it give much insight into their physical significance. However, we are going to show that a boost can be interpreted quite simply in terms of the relative velocity of the charts that it relates. Note that the boosts do not form a group (the product of two symmetric matrices is not symmetric, in general).

Since any boost S is a symmetric, restricted Lorentz transformation, it follows from (B13) that

$$S^0_m = S^m_0 = -\gamma V^m/c = \gamma V'^m/c, \qquad S^0_0 = \gamma. \tag{B14}$$

If K and K' are frames related by S, so that $x' = Sx$, then (B14) implies that the velocity of K' in K is the negative of the velocity of K in K'. We call $\boldsymbol{V}$, which is the velocity of K' in K, the *boost velocity (of S)*.

We can always find a symmetric, restricted, Lorentz transformation S whose elements S^0_m, S^m_0 and S^0_0 satisfy (B14). We simply define

$$\begin{aligned} S^m_n &= \delta_{mn} + (\gamma - 1)V^m V^n |\boldsymbol{V}|^{-2} \\ &= \delta_{mn} + V^m V^n \gamma^2/c^2(\gamma+1). \end{aligned} \tag{B15}$$

One easily checks that (B8) is satisfied and that $\det S > 0$. The next exercise shows that (B15) is *not* the only possible choice for the S^m_n; from now on we define a boost to be a restricted Lorentz transformation for which both (B14) and (B15) hold.

Exercise B4 Find all the symmetric, restricted Lorentz transformations for which (B14) is satisfied. [Suppose that there are two such transformations S and S', and calculate $S^{-1} \circ S' = R$. Show that R is a spatial rotation that leaves $\boldsymbol{V}$ unchanged, and that $R = R^T$. Since $R^2 = RR^T = I$, deduce that R is a rotation through an angle of 0 or π about an axis parallel to $\boldsymbol{V}$.] ■

The coordinate transformation $x' = Sx$ corresponding to the boost (B14) and (B15) is

$$\begin{aligned} x'^0 &= \gamma(x^0 - V^n x^n/c), \\ x'^m &= x^m - \gamma V^m x^0/c + (\gamma - 1)V^m V^n x^n |\boldsymbol{V}|^{-2}. \end{aligned} \tag{B16}$$

In the special case when $V^2 = V^3 = 0$, this reduces to the simple Lorentz transformation (A5). Let us define $b_m = V^m/c$ and $b = V/c$. When V/c is small, one can write $\gamma = 1 + b^2/2 + 3b^4/8 + O_6$, and (B16) becomes

$$\begin{aligned} x'^0 &= (1 + b^2/2 + 3b^4/8 + O_6)x^0 - (1 + b^2/2 + O_4)b^n x^n, \\ x'^m &= x^m - (1 + b^2/2 + O_4)b^m x^0 + (b^m b^n/2 + O_4)x^n. \end{aligned} \tag{B17}$$

Eqs.(B17) are equivalent to VI(1.1).

If $x^0 = 0$ and $x^n V^n = 0$, (B16) implies that $x'^m = x^m$; if $x^0 = 0$ and $x^n = KV^n$, K constant, (B16) implies that $x'^m = \gamma x^m$. For this reason, boosts are sometimes called *restricted Lorentz transformations without rotation of the spatial axes*. The name is misleading: the product of two boosts is not a boost, in general, so that two transformations *without* rotation of the spatial axes can give a transformation *with* rotation of the spatial axes!

We are now ready to prove that any restricted Lorentz matrix can be written as the product of a boost and a spatial rotation. (Alternatively, one says that any restricted Lorentz transformation can be *decomposed* into a boost and a spatial rotation – which accounts for the name of the following theorem.) The behaviour of a system under restricted Lorentz transformations can therefore be deduced from its behaviour under spatial rotations and boosts.

Decomposition theorem Any restricted Lorentz matrix L can be written in the forms

$$L = RS = S'R, \tag{B18}$$

where R is a 3 dimensional rotation, and S and $S' = RSR^T$ are boosts.

Proof Define a boost S by (B15) and the equations $S^0_m = S^m_0 = -\gamma V^m/c$, $S^0_0 = \gamma$ (cf. (B14)), where the V^m are given by (B13). One then has $S^0_\mu = L^0_\mu$. Since $S^{-1} = \eta S^T \eta$, the (0, 0) component of $R = LS^{-1}$ is $L^0_\mu \eta^{\mu\nu} S^\pi_\nu \eta_{\pi 0} = -L^0_\mu \eta^{\mu\nu} S^0_\nu = 1$, and it follows that R is a 3 dimensional rotation. Og $L = RS = RSR^{-1}R$, and $R^{-1} = R^T$ for a 3 dimensional rotation. Obviously $S' = RSR^T$ is symmetric, and it is in fact a boost (use (B14) and (B15)). ■

Inertial charts

At the beginning of this appendix, we announced that we were going to find the most general coordinate transformation between inertial charts – the general Poincaré transformation. The astute reader will have noticed that, although we have indeed found the general Poincaré transformation and discussed its properties, we have never since mentioned inertial charts. To make physical sense of our mathematics, we must give criteria for recognising inertial charts, and show that they are in fact related by Poincaré transformations.

In Appendix A, we gave three criteria that an inertial chart K must satisfy. First, it must be possible to set the clocks fixed in K so that their readings correspond to the time coordinate of K. Second, the Euclidean distance $|\boldsymbol{x} - \boldsymbol{y}| = \left[(x^1 - y^1)^2 + (x^2 - y^2)^2 + (x^3 - y^3)^2\right]^{1/2}$ between any two spatial points $\boldsymbol{x}$ and $\boldsymbol{y}$ of K must be equal to the distance between the corresponding particles as measured by a measuring rod fixed in K. Finally, the speed of light in vacuum as measured in K must always have the value c. We are going to state a generalized form of these criteria in terms of the interval function I, defined by (B1).

We suppose that x and y are the coordinates of spacetime points in a chart K. We call K an *inertial chart* if the following conditions are satisfied for all x and y in $\mathbb{R}^4$.

(i) If an ideal clock reads T_1 at x and T_2 at y and moves with constant velocity in K, then $I(x, y) = -c^2(T_2 - T_1)^2$.

(ii) If a measuring rod has constant proper length L and moves with constant velocity in K, and if x and y are the spacetime coordinates of the ends of the rod, then for any x there is a y such that $L^2 = I(x, y)$ (it is not assumed that $x^0 = y^0$).

(iii) If x and y are on the path of a light signal that travels in vacuum, then $I(x, y) = 0$.

It is clear that the previous criteria are special cases of (i)–(iii).

Since I is invariant under Poincaré transformations, conditions (i)–(iii) are satisfied in any chart K' that is related to an inertial chart K by a Poincaré transformation, and it follows that K' is an inertial chart. (The constant velocities in K, referred to in (i) and (ii), correspond to constant velocities in K'.)

Conversely, if K and K' are inertial charts, conditions (i)–(iii) are satisfied in both of them. Provided that an object with constant velocity in K has constant velocity in K', one can show that I is invariant under the coordinate transformation between K and K' – which is therefore a Poincaré transformation. The proof is not quite immediate because one must show that (ii) holds for any spatially separated x and y if the measuring rod is suitably chosen.

Condition (ii) is rather clumsy. A more elegant formulation is possible if one measures distances by means of clocks and light signals ("radar") rather than measuring rods. We suppose that x and y are the coordinates in K of two spacetime points and that $I(x, y) > 0$. The point x lies on the path of a clock that moves with constant velocity in K at a speed less than the speed of light. The clock emits a light signal at the point w, which is reflected without delay at y, and returns to the clock at the point z. The light signals travel in vacuum, so that $I(w, y) = I(y, z) = 0$. The points w, x, and z lie on the straight-line path of the clock and satisfy $z - x = \lambda(x - w)$, where $\lambda > 0$.

To simplify the notation, we write $I(x, y) = \eta_{\mu\nu}(x^\mu - y^\mu)(x^\nu - y^\nu) = (x - y) \cdot (x - y)$, $\eta_{\mu\nu}(x^\mu - y^\mu)(w^\nu - z^\nu) = (x - y) \cdot (w - z)$, etc. We then have

$$\begin{aligned} I(w, y) &= (y - x + x - w) \cdot (y - x + x - w) \\ &= I(x, y) + I(w, x) + 2(x - w) \cdot (y - x) = 0, \\ I(z, y) &= I(x, y) + I(z, x) + 2(x - z) \cdot (y - x) = 0. \end{aligned}$$

We put $x - z = -\lambda(x - w)$ in the last equation, and note that $I(z, x) = \lambda^2 I(w, x)$. We then multiply the previous equation by λ and add to get $(\lambda + 1)I(x, y) + (\lambda + \lambda^2)I(w, x) = 0$. Since $\lambda + 1 \neq 0$, og $I(x, y) = -\lambda I(w, x)$, and hence

$$I(x, y)^2 = \lambda^2 I(w, x)^2 = I(w, x) I(z, x). \tag{B19}$$

If T_1, T_2, and T_3 are the times read by the clock at w, x, and z, respectively, then it follows from (i) and (B19) that

$$I(x,y) = c^2(T_2 - T_1)(T_3 - T_2). \tag{B20}$$

We replace (ii) by the statement that the times in the radar measurement are related to the interval $I(x,y)$ by (B20). Because of the invariance of I, an exactly similar equation holds in any chart K' that is related to K by a Poincaré transformation (the constant velocity of the clock in K corresponds to a constant velocity in K'). Conversely, if K and K' are inertial charts, then (B20) holds in K and an exactly similar equation in K', and one has $I(x,y) = I(x',y')$ for all x, y such that $I(x,y) > 0$. For more details, see [Synge 58], Chapter 1.

The Principle of Relativity

For a deeper discussion of the significance of inertial charts, the reader must go elsewhere. We note only that special relativity assumes that the laws of nature are the same in all inertial charts. This is sometimes called the *Principle of Relativity.* Because it is hard to say precisely what one means by a *law of nature*, it is better to state the Principle of Relativity in one of the following ways.

(1) The description of a possible set of events in an inertial chart is the description of a possible set of events in any other inertial chart. (The sets of events are in general different!)

(2) The set of descriptions of possible sets of events is the same in all inertial charts.

The statements (1) and (2) are of course equivalent. To understand more clearly what (1) means, imagine that you have a description of a possible set of events in a file with the label *This description applies to the inertial chart K.* If you lose the label, (1) says that you cannot find out, from the description itself, to which inertial chart the file refers.

Notes

[1]This is the correct way to state it, because x^0 (like the x^m) has the dimension of length. However, people often speak of x^0 as the time coordinate, and we shall follow their bad example.

[2]We have used the fact that the left-inverse of a matrix is equal to the right-inverse. To prove this, let A, B, and C be $N \times N$ matrices such that $AB = CA = I$, where I is the $N \times N$ unit matrix. Then $CAB = IB = B$, and $CAB = CI = C$, and hence $B = C$.

[3]We are not going to use any significant theorems of group theory. You should know, however, that a group is a set of elements on which an associative multiplication is defined, that there is a unit element, and that each element has an inverse that is an element of the group. Two groups are isomorphic if there is a one-to-one correspondence between their elements that is compatible with the group products. More precisely, the groups G and H are isomorphic if there is a bijection $T : G \to H$ such that $T(gg') = T(g)T(g')$ for all $g, g' \in G$. Note that gg' is the product of g and g' as elements of G, and that $T(g)T(g')$ is the product of $T(g)$ and $T(g')$ as elements of H. We say that T is the isomorphism of G onto H. If T is a surjection (onto but not necessarily one-to-one) then G and H are homomorphic, and we say that T is the homomorphism of G onto H.

[4]Read *resp.* as *respectively*. The sentence is equivalent to "A Lorentz matrix is *proper* or *improper* if $\det L = 1$ or $\det L = -1$, respectively".

C Postnewtonian Potentials

In Chapter VI we expressed the postnewtonian metric in terms of postnewtonian potentials, the simplest of which are the functions χ and U defined by VI(2.6) and V(2.4):

$$\chi(x) = -kc^2 \int \rho(x^0, \boldsymbol{y})|\boldsymbol{x} - \boldsymbol{y}|\, d^3y, \tag{C1}$$

$$U(x) = kc^2 \int \rho(x^0, \boldsymbol{y})|\boldsymbol{x} - \boldsymbol{y}|^{-1}\, d^3y. \tag{C2}$$

As before, it is understood that integrals are over all values of the integration variables unless stated otherwise, and that $(x) = (x^0, \boldsymbol{x})$. We recall that a function is *smooth*, or C^∞, if its partial derivatives of every order exist and are continuous, and that the *support* of a function is the set of points at which the function does not vanish. The function ρ (the invariant mass density) is assumed to be smooth, and to have spatially bounded support: for each x^0 there is a constant $A > 0$ such that $\rho(x^0, \boldsymbol{x}) = 0$ for all $|\boldsymbol{x}| > A$.

Derivatives of χ

In calculating partial derivatives of potentials such as χ and U, one can sometimes use standard results of elementary calculus. When the integrand is sufficiently regular, for example, the integral and the partial derivative commute, and one can "differentiate under the integral sign" in the usual manner. With less regular integrands this is not always possible, and one has to consider other methods.

For brevity, let us write $r = |\boldsymbol{x} - \boldsymbol{y}| = [(x^m - y^m)(x^m - y^m)]^{1/2}$ and $\partial_m = \partial/\partial x^m$, and define $K_m = (x^m - y^m)r^{-1}$. The K_m are the components of a unit vector ($K_m K_m = 1$), and og

$$\begin{aligned} \partial_m r &= (x^m - y^m)r^{-1}, \\ \partial_n \partial_m r &= \delta_{mn} r^{-1} - (x^m - y^m)(x^n - y^n)r^{-3} = (\delta_{mn} - K_m K_n)r^{-1}, \end{aligned} \tag{C3}$$

$$\partial_m r^{-1} = -(x^m - y^m)r^{-3} = -K_m r^{-2},$$
$$\partial_n\partial_m r^{-1} = -\delta_{mn}r^{-3} + 3(x^m - y^m)(x^n - y^n)r^{-5} = (-\delta_{mn} + 3K_mK_n)r^{-3}, \tag{C4}$$
$$\partial_m\partial_m r^{-1} = \nabla^2 r^{-1} = 0. \tag{C5}$$

All these equations are valid for $r > 0$; the partial derivatives are not defined at $r = 0$.

The $\partial_m r$ are discontinuous but bounded as $r \to 0$; the $\partial_n\partial_m r$, $\partial_m r^{-1}$ and $\partial_n\partial_m r^{-1}$, diverge like r^{-1}, r^{-2}, and r^{-3}, respectively. If one ignores this difficulty, and simply differentiates (C1) formally, assuming that partial differentiation and integration commute, one finds from (C3) that

$$\begin{aligned}\partial_n\partial_m\chi(x) &= -kc^2\int \rho(x^0, \boldsymbol{y})(\delta_{mn} - K_mK_n)r^{-1}\,d^3y \\ &= -\delta_{mn}U(x) + \mathcal{U}_{mn}(x),\end{aligned} \tag{C6}$$

and hence

$$\partial_m\partial_m\chi = \nabla^2\chi = -2U, \tag{C7}$$

where (cf. VI(1.8))

$$\mathcal{U}_{mn}(x) = kc^2\int \rho(x^0, \boldsymbol{y})K_mK_n|\boldsymbol{x} - \boldsymbol{y}|^{-1}\,d^3y. \tag{C8}$$

The first objection to this procedure is that the integral $\int \rho(x^0, \boldsymbol{y})\partial_n\partial_m r\,d^3y$ is not well defined because the integrand does not exist at the origin. One can define the integral to be the limit as $b \to 0$ of the expression

$$\int_{r>b} \rho(x^0, \boldsymbol{y})\partial_n\partial_m r\,d^3y = \int_{r>b} \rho(x^0, \boldsymbol{y})(\delta_{mn} - K_mK_n)r^{-1}\,d^3y, \tag{C9}$$

where the integration is over all $\boldsymbol{y}$ such that $r = |\boldsymbol{x} - \boldsymbol{y}| > \text{b}$. The 3 dimensional volume element expressed in terms of the polar coordinates r, θ, ϕ, is $d^3y = r^2\sin\theta dr d\theta d\phi$, and $r^2\partial_n\partial_m r$ is continuous at $r = 0$. Since ρ is continuous, it is easy to see that the limit exists.

One can prove that (C6), with the integral interpreted as in (C9), is correct. Similarly, one proves that since $\partial_0\rho$ exists and is continuous,

$$\begin{aligned}\partial_0\chi(x) &= -kc^2\int \partial_0\rho(x^0, \boldsymbol{y})|\boldsymbol{x} - \boldsymbol{y}|\,d^3y \\ &= kc\int \rho V^m(x^0, \boldsymbol{y})K_m\,d^3y + LO_5,\end{aligned} \tag{C10}$$

where we have used the continuity equation V(5.5) in the form $c\partial_0\rho = -\partial_m(\rho V^m) + \rho c L^{-1} O_3$ (recall that $U = O_2$ and $\chi = L^2 O_2$, and hence that $\partial_0\chi = LO_3$). Differentiating (C.10), og

$$\partial_0\partial_m\chi = \mathcal{V}_m - \mathcal{W}_m + O_5, \tag{C11}$$

where

$$\mathcal{V}_m(x) = kc\int \rho V^m(x^0, \boldsymbol{y})|\boldsymbol{x} - \boldsymbol{y}|^{-1}\, d^3y, \tag{C12}$$

$$\mathcal{W}_m(x) = kc\int \rho V^n(x^0, \boldsymbol{y}) K_n K_m |\boldsymbol{x} - \boldsymbol{y}|^{-1}\, d^3y. \tag{C13}$$

To calculate $\partial_0^2\chi$, we use the Euler equation VII(2.3) (with the appropriate postnewtonian error term $\rho c^2 L^{-1} O_4$). Og

$$\partial_0^2\chi = -2\Phi_4 - \Phi_1 + \mathcal{A} - \xi + O_6, \tag{C14}$$

where

$$\mathcal{A}(x) = k\int \rho(x^0, \boldsymbol{y})\big[V^m(x^0, \boldsymbol{y})(x^m - y^m)\big]^2 |\boldsymbol{x} - \boldsymbol{y}|^{-3}\, d^3y, \tag{C15}$$

$$\xi(x) = k^2 c^4 \int \rho(x^0, \boldsymbol{y})\rho(x^0, \boldsymbol{z})(y^m - z^m)(x^m - y^m)|\boldsymbol{y} - \boldsymbol{z}|^{-3}|\boldsymbol{x} - \boldsymbol{y}|^{-1}\, d^3y d^3z. \tag{C16}$$

The potentials Φ_α, $\alpha \in \{1, 2, 3, 4\}$, are defined as in VI(1.13)–VI(1.16):

$$\Phi_1(x) = k\int \rho|\boldsymbol{V}|^2(x^0, \boldsymbol{y})|\boldsymbol{x} - \boldsymbol{y}|^{-1}\, d^3y, \tag{C17}$$

$$\Phi_2(x) = kc^2\int \rho U(x^0, \boldsymbol{y})|\boldsymbol{x} - \boldsymbol{y}|^{-1}\, d^3y, \tag{C18}$$

$$\Phi_3(x) = kc^2\int \rho\Pi(x^0, \boldsymbol{y})|\boldsymbol{x} - \boldsymbol{y}|^{-1}\, d^3y, \tag{C19}$$

$$\Phi_4(x) = k\int p(x^0, \boldsymbol{y})|\boldsymbol{x} - \boldsymbol{y}|^{-1}\, d^3y. \tag{C20}$$

Difficulties with derivatives of U

The calculation of the partial derivatives of U is not quite so straightforward. If one follows the same procedure as before, and simply differentiates (C2) assuming that partial differentiation and integration commute, one finds from (C4) that $\partial_n\partial_m U$ is given by

$$kc^2\int \rho(x^0, \boldsymbol{y})\partial_n\partial_m r^{-1}\, d^3y = kc^2\int \rho(x^0, \boldsymbol{y})(-\delta_{mn} + 3K_m K_n)r^{-3}\, d^3y.$$

The integral on the right-hand side is defined to be the limit as $b \to 0$ of the expression

$$kc^2 \int_{r>b} \rho(x^0, \boldsymbol{y})(-\delta_{mn} + 3K_m K_n) r^{-3}\, d^3y, \tag{C21}$$

which exists provided that ρ is smooth at $\boldsymbol{x} = \boldsymbol{y}$ (so that $\rho(x^0, \boldsymbol{y}) = \rho(x^0, \boldsymbol{x}) +$ terms of order b, when $r \leq b$). Unfortunately, this expression for $\partial_n \partial_m U$ is not correct. It implies, for example, that $\partial_m \partial_m U = \nabla^2 U = 0$ for any choice of ρ!

Generalized functions

One can find the correct expression for the derivatives $\partial_n \partial_m U$ by the methods of classical analysis. Some care is necessary, because r^{-1} and its derivatives do not exist at $r = 0$. It is more elegant to regard r^{-1} as a *generalized function*, whose partial derivatives always exist. We shall give a brief outline of the theory of generalized functions – just sufficient to enable the reader to use it with confidence. For more complete treatments, see [Lighthill 58] or [Gelfand 64].

There are several ways of defining generalized functions. Perhaps the simplest is to regard a generalized function as a linear function on a set S of functions on $\mathbb{R}^N$. The elements of S are called *test functions* or *good functions*. We choose them to be smooth and of bounded support. (More generally, as in [Lighthill 58], they may be chosen to be rapidly decreasing at large distances from the origin.) A generalized function ζ assigns a real number ζf to each test function, and ζ is linear: it satisfies $\zeta(bf + cg) = b\zeta f + c\zeta g$, for all test functions f and g, and all real numbers b and c.

An example of a generalized function is the *Dirac delta function* δ^N, which assigns to each test function on $\mathbb{R}^N$ its value at the origin: $\delta^N f = f(0)$. More generally, one may define the generalized function δ_a^N by $\delta_a^N f = f(a)$, for any $a \in \mathbb{R}^N$. If $N = 1$, one may write $\delta^1 = \delta$ and $\delta_a^1 = \delta_a$, just as one writes $\mathbb{R}^1 = \mathbb{R}$. We adopt the notation

$$\delta_a^N f = \int \delta^N(x - a) f(x)\, d^N x = f(a). \tag{C22}$$

Sometimes, deplorably, we may write $\delta^N(x - a)$ instead of δ_a^N; the reader who is accustomed to speak of the function $f(x)$ will not shudder.

Any ordinary, integrable function F can be interpeted as a generalized function ${}^gF$. One simply assigns to any test function f the real number ${}^gFf = \int F(x) f(x)\, d^N x$ (the linearity condition is obviously satisfied). The ordinary functions are therefore a subset of the generalized functions, in much the same way that the rational numbers are a subset of the reals. Usually we shall not distinguish notationally between the ordinary function F and the generalized function ${}^gF$ – we write both of them as F. However, the product Ff has different meanings in the two interpretations!

The product Ff of a smooth function F and a test function f is a test function (it is C^∞ because F and f are C^∞, and of bounded support because f is of bounded support). We can therefore define the product of the smooth function F and the generalized function ζ to be the generalized function $F\zeta$ such that $(F\zeta)f = \zeta(Ff)$ for every test function f. (Note that $F\zeta \neq \zeta$F.) It is trivial to show that if G is another smooth function, then $(F+G)\zeta = F\zeta + G\zeta$ and $F(G\zeta) = (FG)\zeta = G(F\zeta)$. The product of two generalized functions is not defined in general.

If F is a smooth function and f is a test function, og $(\partial_m F)f = \partial_m(Ff) - F\partial_m f$. Since f is of bounded support, the $\partial_m(Ff)$ term vanishes when one integrates over all x, and og

$$\int (\partial_m F) f(x)\, d^N x = -\int F(\partial_m f)(x)\, d^N x.$$

This motivates the definition of the partial derivative $\partial_m\zeta$ of the generalized function ζ as the generalized function such that

$$(\partial_m \zeta) f = -\zeta(\partial_m f), \tag{C23}$$

for all test functions f. Note that the partial derivative of a generalized function always exists. It is easy to show that $\partial_m(F\zeta) = (\partial_m F)\zeta + F\partial_m\zeta$ for any smooth function F and generalized function ζ. Repeated application of (C23) gives $(\partial_m\partial_n\zeta)f = \zeta(\partial_m\partial_n f), (\partial_m\partial_n\partial_p\zeta)f = -\zeta(\partial_m\partial_n\partial_p f)$, etc. In the case $N = 1$, we write the derivative of ζ as $D\zeta$, just as for ordinary functions.

As an example, we consider the *Heaviside step function*, which is defined as an ordinary function on $\mathbb{R}$ by $H(x) = 0$ if $x < 0$ and $H(x) = 1$ if $x \geq 0$. One defines H as a generalized function by

$$Hf = \int Hf(x)dx = \int_0^\infty f(x)dx, \tag{C24}$$

for any test function f. The derivative DH of H is given by

$$(DH)f = -\int HDf(x)\, dx = -\int_0^\infty Df(x)\, dx = f(0),$$

and hence $DH = \delta$: the derivative of the Heaviside step function is the Dirac delta function. Similarly, if one defines the ordinary function sgn by sgn $x = -1$ if $x < 0$, sgn $0 = 0$, and sgn $x = 1$ if $x > 0$, og $D\,\mathrm{sgn} = 2\delta$.

Translate of a generalized function

We have introduced the Dirac delta function δ^N that assigns to any test function f its value at the origin, and the generalized function δ_a^N that assigns to f its value at the point a. We call δ_a^N the *translate of* δ^N *by* a. Similarly, for any generalized function ζ on $\mathbb{R}^N$ and any $a \in \mathbb{R}^N$, we can define a generalized function ζ_a on $\mathbb{R}^N$, the *translate of* ζ *by* a.

First, for any ordinary function F on $\mathbb{R}^N$, we define a function F_a by $F_a(x) = F(x-a)$ for all x (i.e. the value of F at x is the value of F_a at $x+a$). If F is smooth, so is F_a; if F is a test function, so is F_a. We then define ζ_a by $\zeta_a f = \zeta f_{-a}$ for every test function f. In the case of the Dirac delta function, this becomes $\delta_a^N f = \delta^N f_{-a} = f_{-a}(0) = f(0-(-a)) = f(a)$, in agreement with the previous definition.

As before, we define ${}^gF$ to be the generalized function that corresponds to the ordinary function F. The translate of ${}^gF$ by a is given by

$$\begin{aligned} {}^gF_a f = {}^gF f_{-a} &= \int F(x) f_{-a}(x)\, d^N x \\ &= \int F(x) f(x+a)\, d^N x \\ &= \int F(x-a) f(x)\, d^N x, \end{aligned}$$

for any test function f. Note the resemblance between this equation and (C22).

Since $a = (a_1, \ldots, a_N) \in \mathbb{R}^N$, the generalized function ζ_a (the translate of ζ by a) depends on the parameters a_m. We define the *derivative of* ζ_a *with respect to* a_m to be the generalized function $(\partial/\partial a_m)\zeta_a$ such that $[(\partial/\partial a_m)\zeta_a]f = (\partial/\partial a_m)(\zeta_a f)$ for all test functions f. (It is assumed that the test functions do not depend on a.)

The derivative $(\partial/\partial a_m)\zeta_a$ is simply related to $\partial_m \zeta_a$. We define $b \in \mathbb{R}^N$ by $b_i = a_i + \lambda\delta_{im}$, where λ is real. For any test function f that does not depend on a, og $(\zeta_b - \zeta_a)f = \zeta(f_{-b} - f_{-a})$. From a theorem on C^∞ functions, $(f_{-b} - f_{-a})(x) = f(x+b) - f(x+a) = \lambda\partial_m f(x+a) + \lambda^2 g_{(m)}(x, \lambda)$, where $g_{(m)}(x,\lambda)$ is bounded as $\lambda \to 0$. We have $\partial_m f(x+a) = \partial_m f_{-a}(x)$, and hence $(\zeta_b - \zeta_a)f = \lambda\zeta\partial_m f_{-a} + \lambda^2 \zeta g_{(m)}(\cdot\,,\lambda)$. It follows from the definition of the derivative of an ordinary function that

$$\begin{aligned} [(\partial/\partial a_m)\zeta_a]f &= (\partial/\partial a_m)(\zeta_a f) \\ &= \lim{}_{\lambda\to 0}\lambda^{-1}(\zeta_b - \zeta_a)f = \zeta(\partial_m f_{-a}) = -(\partial_m \zeta) f_{-a}, \end{aligned}$$

where we have used (C23).

Exercise C1 Show that $\partial_m f_{-a} = (\partial_m f)_{-a}$, and hence that $(\partial_m \zeta) f_{-a} = (\partial_m \zeta_a) f$, and that

$$(\partial/\partial a_m)\zeta_a = -\partial_m \zeta_a. \ \blacksquare$$

Exercise C2 In the special case when $\zeta = {}^g F$, where F is an ordinary differentiable function, show that the result of Exercise C1 is equivalent to the equation $(\partial/\partial a_m)F(x-a) = -\partial_m F(x-a)$. ∎

Exercise C3 We have been assuming that the test functions f are independent of a. If they do depend on a, *and if* $(\partial/\partial a_m)f$ *is a test function*, we define

$$[(\partial/\partial a_m)\zeta_a]f = (\partial/\partial a_m)(\zeta_a f) - \zeta_a(\partial/\partial a_m)f.$$

Is this definition compatible with the result of Exercise C1? ∎

Derivatives of R

We next consider generalized functions on $\mathbb{R}^3$ (the special case $N = 3$). We write $r = |\boldsymbol{x}| = [(x^m x^m]^{1/2}$ (the previous definition with $\boldsymbol{y} = 0$) and define the ordinary function R^{-1} by $R^{-1}(\boldsymbol{x}) = r^{-1}$ for $r > 0$. We define R^{-1} as a generalized function by

$$R^{-1}f = \int R^{-1} f \, d^3x = \int r^{-1} f(x)\, d^3x. \tag{C25}$$

To show that this is well defined for every test function f, we write the integral in terms of polar coordinates r, θ, ϕ, for which the volume element is $r^2 \sin\theta \, dr d\theta d\phi$. Similarly, for any positive integer M, we define an ordinary function R^{-M} by $R^{-M}(\boldsymbol{x}) = r^{-M}$ for $r > 0$. However, the integral $\int r^{-M} f(\boldsymbol{x})\, d^3x$ is *not* defined for all test functions when $M > 2$, and we cannot define the generalized function R^{-M} in the manner of (C25).

Note that we have, very properly, distinguished the ordinary function R^{-M} from its value $R^{-M}(\boldsymbol{x}) = r^{-M}$. When applying the theory, one often ignores this distinction, and writes the (ordinary or generalized) function R^{-M} as r^{-M}.

The partial derivatives of the generalized function R^{-1} are given by (C23) and (C25):

$$(\partial_m R^{-1})f = -R^{-1}\partial_m f = -\int r^{-1}\partial_m f(x)\, d^3x, \tag{C26}$$

$$(\partial_m \partial_n R^{-1})f = R^{-1}\partial_m\partial_n f = \int r^{-1}\partial_m\partial_n f(x)\, d^3x, \tag{C27}$$

etc. Writing the integral in (C27) in terms of polar coordinates, we see that the integrand is of order r when r is small, and og

$$(\partial_m \partial_n R^{-1})f = \lim_{b\to 0} \int_{r>b} r^{-1} \partial_m \partial_n f(x)\, d^3\boldsymbol{x}. \tag{C28}$$

We have $r^{-1}\partial_m\partial_n f = \partial_m(r^{-1}\partial_n f) - (\partial_m r^{-1})\partial_n f = \partial_p(\delta_{pm} r^{-1}\partial_n f) - \partial_p(\delta_{pn}(\partial_m r^{-1})f) + (\partial_n\partial_m r^{-1})f$. From (C4) og $\partial_m r^{-1} = -x^m r^{-3}$, $\partial_n\partial_m r^{-1} = -\delta_{mn} r^{-3} + 3x^m x^n r^{-5}$. We substitute into (C28) and evaluate the first two terms by means of the divergence theorem. (Apply the theorem to the region between $r = b$ and $r = B$ and then let $B \to \infty$; note that f is of bounded support so that the integrals over $r = B$ vanish in the limit; remember that the unit vector in the radial direction is $\boldsymbol{x} r^{-1}$.) Og

$$\begin{aligned}(\partial_m \partial_n R^{-1})f = \lim_{b\to 0}\Big\{ &-\int_{r=b} x^p r^{-1} \delta_{pm} r^{-1} \partial_n f(\boldsymbol{x}) r^2 \sin\theta\, dr\, d\theta\, d\phi \\ &-\int_{r=b} x^p r^{-1} \delta_{pn} x^m r^{-3} f(\boldsymbol{x}) r^2 \sin\theta\, dr\, d\theta\, d\phi \\ &+\int_{r>b} (-\delta_{mn} r^{-3} + 3x^m x^n r^{-5}) f(\boldsymbol{x}) r^2 \sin\theta\, dr\, d\theta\, d\phi\Big\}. \end{aligned} \tag{C29}$$

Since f is smooth, og $f(\boldsymbol{x}) = f(0) + O(b)$ and $\partial_n f(\boldsymbol{x}) = \partial_n f(0) + O(b)$ when $r = b$. We put $\boldsymbol{x} r^{-1} = (\cos\theta, \sin\theta\cos\phi, \sin\theta\sin\phi)$, and find that the first integral in (C29) vanishes, while the second is $-(4/3)\pi\delta_{mn} f(0) + O(b)$. The limit of the last integral as $b \to 0$ exists in spite of the r^{-3} factors. Eq.(C29) holds for every test function f. We have therefore proved that $\partial_m\partial_n R^{-1} = G_{mn} - (4\pi/3)\delta_{mn}\delta^3$, where $G_{mn}(\boldsymbol{x}) = -\delta_{mn} r^{-3} + 3x^m x^n r^{-5}$ for $r > 0$.

We can replace the origin by the point $\boldsymbol{y}$ in the preceding calculations, and R by its translate $R_{\boldsymbol{y}}$ (recall that $R_{\boldsymbol{y}}(\boldsymbol{x}) = R(\boldsymbol{x} - \boldsymbol{y})$). We write $r = |\boldsymbol{x} - \boldsymbol{y}|$ and $K_m = (x^m - y^m)r^{-1}$ as before, define $G_{\boldsymbol{y}mn}$ by $G_{\boldsymbol{y}mn}(\boldsymbol{x}) = (-\delta_{mn} + 3K_m K_n)r^{-3}$ for $r > 0$, and get

$$\partial_m\partial_n R_{\boldsymbol{y}}^{-1} = G_{\boldsymbol{y}mn} - (4/3)\pi\delta_{mn}\delta^3_{\boldsymbol{y}}, \tag{C30}$$

and hence

$$\partial_m\partial_m R_{\boldsymbol{y}}^{-1} = \nabla^2 R_{\boldsymbol{y}}^{-1} = -4\pi\delta^3_{\boldsymbol{y}}. \tag{C31}$$

The generalized function $R_{\boldsymbol{y}}^{-1}$ is commonly written as r^{-1} or $|\boldsymbol{x} - \boldsymbol{y}|^{-1}$ in equations such as (C30) and (C31), and $\delta^3_{\boldsymbol{y}}$ is written as $\delta^3(\boldsymbol{x} - \boldsymbol{y})$.

Derivatives of postnewtonian potentials

We are now at last able to calculate the partial derivatives of the potential U and the other postnewtonian potentials. Since the density ρ is assumed to be smooth and of spatially bounded support, we can regard it as a test function. Eq.(C2) can then be written as $U = kc^2 R_{\boldsymbol{x}}^{-1}\rho$. (The integration variables in (C2) are the y^m, so the roles of $\boldsymbol{x}$ and $\boldsymbol{y}$ are switched with respect to those in (C30).) Differentiating with respect to the parameter x^s and using the definition of $(\partial/\partial x^s)R_{\boldsymbol{x}}^{-1}$ and the result of Exercise C1, og $(\partial/\partial x^s)U = kc^2(\partial/\partial x^s)R_{\boldsymbol{x}}^{-1}\rho = -kc^2(\partial_s R_{\boldsymbol{x}}^{-1})\rho$, and similarly $(\partial/\partial x^r)(\partial/\partial x^s)U = kc^2(\partial_r\partial_s R_{\boldsymbol{x}}^{-1})\rho$. We substitute from (C30) (with $\boldsymbol{x}$ and $\boldsymbol{y}$ exchanged), write $(\partial/\partial x^r)(\partial/\partial x^s)U(x) = \partial_r\partial_s U(x)$, and find that

$$\begin{aligned}\partial_r\partial_s U(x) &= kc^2\int \rho(x^0,\boldsymbol{y})\partial_r\partial_s|\boldsymbol{x}-\boldsymbol{y}|^{-1}\,d^3y \\ &= kc^2\int \rho(x^0,\boldsymbol{y})(-\delta_{rs}+3K_rK_s)|\boldsymbol{x}-\boldsymbol{y}|^{-3}\,d^3y \\ &\quad -(4/3)\pi kc^2\delta_{rs}\rho(x).\end{aligned} \tag{C32}$$

When $r = s$, (C32) becomes the Poisson equation

$$\partial_m\partial_m U = \nabla^2 U = -4\pi kc^2\rho, \tag{C33}$$

which can also be shown directly from (C31).

In a similar manner, but using the continuity equation V(5.5), og

$$\partial_0 U(x) = -kc\int \rho V^r(x^0,\boldsymbol{y})\partial_r|\boldsymbol{x}-\boldsymbol{y}|^{-1}\,d^3y + L^{-1}O_5, \tag{C34}$$

$$\begin{aligned}\partial_s\partial_0 U(x) &= -kc\int \rho V^r(x^0,\boldsymbol{y})(-\delta_{rs}+3K_rK_s)|\boldsymbol{x}-\boldsymbol{y}|^{-3}\,d^3y \\ &\quad +(4/3)\pi kc\rho V^s(x) + L^{-2}O_5.\end{aligned} \tag{C35}$$

Differentiating (C7), og $-2\partial_0^2 U = \partial_m\partial_m\partial_0^2\chi$. Substituting from (C14) and using (C31) gives

$$\partial_0^2 U = -4\pi kp - 2\pi k\rho|\boldsymbol{V}|^2 - (1/2)\partial_m\partial_m\mathcal{A} + (1/2)\partial_m\partial_m\xi + L^{-2}O_6. \tag{C36}$$

It follows immediately from (C31) and (C12) that

$$\partial_r\partial_r\mathcal{V}_m = \nabla^2\mathcal{V}_m = -4\pi kc\rho V^m. \tag{C37}$$

Similarly, for the potentials Φ_α, $\alpha\in\{1,2,3,4\}$, defined by (C17)–(C20) og

$$\partial_r\partial_r\Phi_\alpha = \nabla^2\Phi_\alpha = -4\pi ks_\alpha, \tag{C38}$$

where $s_1 = \rho|\boldsymbol{V}|^2$, $s_2 = c^2\rho U$, $s_3 = c^2\rho\Pi$, $s_4 = p$. To find similar expressions for $\nabla^2\mathcal{W}_m$ and $\nabla^2\mathcal{U}_{mn}$, we must calculate

$$\begin{aligned}
&\partial_r\partial_r(K_nK_m|\boldsymbol{x}-\boldsymbol{y}|^{-1})\\
&\quad=\partial_r\partial_r\Big[(x^n-y^n)(x^m-y^m)|\boldsymbol{x}-\boldsymbol{y}|^{-3}\Big]\\
&\quad=-\partial_r\partial_r\Big[(x^n-y^n)\partial_m|\boldsymbol{x}-\boldsymbol{y}|^{-1}\Big]\\
&\quad=-2\partial_n\partial_m|\boldsymbol{x}-\boldsymbol{y}|^{-1}-(x^n-y^n)\partial_m\partial_r\partial_r|\boldsymbol{x}-\boldsymbol{y}|^{-1}\\
&\quad=2(\delta_{mn}-3K_mK_n)|\boldsymbol{x}-\boldsymbol{y}|^{-3}+(8\pi/3)\delta_{mn}\delta^3(\boldsymbol{x}-\boldsymbol{y})\\
&\qquad\quad+4\pi(x^n-y^n)\partial_m\delta^3(\boldsymbol{x}-\boldsymbol{y})\\
&\quad=2(\delta_{mn}-3K_mK_n)|\boldsymbol{x}-\boldsymbol{y}|^{-3}-(4\pi/3)\delta_{mn}\delta^3(\boldsymbol{x}-\boldsymbol{y}),
\end{aligned}\tag{C39}$$

from (C30) and (C31). In the last step, we have used the fact that $x^n\delta^3=0$, which implies that $\partial_m(x^n\delta^3)=\delta_{mn}\delta^3+x^n\partial_m\delta^3=0$, and hence that $(x^n-y^n)\partial_m\delta^3(\boldsymbol{x}-\boldsymbol{y})=-\delta_{mn}\delta^3(\boldsymbol{x}-\boldsymbol{y})$.

Eqs.(C39), (C13), and (C8) give

$$\begin{aligned}
\nabla^2\mathcal{W}_m(x)&=2kc\int\rho V^n(x^0,\boldsymbol{y})(\delta_{mn}-3K_mK_n)|\boldsymbol{x}-\boldsymbol{y}|^{-3}\,d^3y\\
&\qquad-(4/3)\pi kc\rho V^m(x),
\end{aligned}\tag{C40}$$

$$\begin{aligned}
\nabla^2\mathcal{U}_{mn}(x)&=2kc^2\int\rho(x^0,\boldsymbol{y})(\delta_{mn}-3K_mK_n)|\boldsymbol{x}-\boldsymbol{y}|^{-3}\,d^3y\\
&\qquad-(4/3)\pi kc^2\delta_{mn}\rho(x).
\end{aligned}\tag{C41}$$

Eqs.(C32) and (C35) can therefore be rewritten as

$$2\partial_r\partial_sU=-\nabla^2\mathcal{U}_{rs}+\delta_{rs}\nabla^2U,\tag{C42}$$

$$2\partial_r\partial_0U=\nabla^2\mathcal{W}_r-\nabla^2\mathcal{V}_r+L^{-2}O_5.\tag{C43}$$

D Covariant Divergence and Covariant Derivative

In special relativity, the divergence of a vector field a with components a^μ in an inertial chart y is defined to be $a^\mu{}_{,\mu} = \partial a^\mu / \partial y^\mu$. The divergence is a Poincaré invariant: if a has components a'^μ in any other inertial chart y', then $a'^\mu{}_{,\mu} = \partial a'^\mu / \partial y'^\mu = a^\mu{}_{,\mu}$. One sometimes calls $a^\mu{}_{,\mu}$ the *ordinary divergence* of a.

If y or y' is not an inertial chart, then $a^\mu{}_{,\mu} \neq a'^\mu{}_{,\mu}$ in general. We are going to find an invariant quantity, the *covariant divergence* of a, that has the same form for all charts, inertial or not. The covariant divergence reduces to the ordinary divergence in an inertial chart; it is what we need to write the equation of continuity in the covariant form V(4.11), for example. In a similar manner, we shall define the covariant divergence of a second-order tensor field to be a vector (or covector) field.

The formal difference between ordinary and covariant divergences is that the partial derivative is replaced by a new kind of derivative: the *covariant derivative*. We shall list the properties of the covariant derivative that are needed in Chapter IX. These properties, and the expressions for the covariant divergence, are not restricted to special relativity, but are valid for theories with a general spacetime metric.

Algebraic preliminaries

As usual, we assume that lower-case Greek indices have the range $\{0,1,2,3\}$ and obey the summation convention. The alternating symbol $\epsilon_{\mu\nu\lambda\pi}$ is defined to be zero if μ, ν, λ, π are not distinct, to be 1 if (μ, ν, λ, π) is an even permutation of (0, 1, 2, 3), and to be -1 if it is an odd permutation.

In any chart y, the components of the spacetime metric are symmetric: $g_{\mu\nu} = g_{\nu\mu}$. We define the determinant of the matrix with elements $g_{\mu\nu}$ to be $G = \det(g_{\mu\nu})$. Since the metric is nondegenerate, G vanishes nowhere. From the definition of the determinant, og

$$G = \epsilon_{\mu\nu\lambda\pi}\, g_{0\mu}\, g_{1\nu}\, g_{2\lambda}\, g_{3\pi}. \tag{D1}$$

It follows that

$$\epsilon_{\rho\sigma\tau\theta} G = \epsilon_{\mu\nu\lambda\pi}\, g_{\rho\mu}\, g_{\sigma\nu}\, g_{\tau\lambda}\, g_{\theta\pi}, \tag{D2}$$

and multiplying by $\epsilon_{\rho\sigma\tau\theta}$ og

$$G = (1/24)\epsilon_{\rho\sigma\tau\theta}\, \epsilon_{\mu\nu\lambda\pi}\, g_{\rho\mu}\, g_{\sigma\nu}\, g_{\tau\lambda}\, g_{\theta\pi}. \tag{D3}$$

The functions $g^{\rho\mu}$ are defined by

$$g^{\rho\mu} = (1/6G)\epsilon_{\rho\sigma\tau\theta}\, \epsilon_{\mu\nu\lambda\pi}\, g_{\sigma\nu}\, g_{\tau\lambda}\, g_{\theta\pi}. \tag{D4}$$

From (D2) og

$$\begin{aligned} g^{\rho\mu} g_{\kappa\mu} &= (1/6G)\epsilon_{\rho\sigma\tau\theta}\, \epsilon_{\mu\nu\lambda\pi}\, g_{\kappa\mu}\, g_{\sigma\nu}\, g_{\tau\lambda}\, g_{\theta\pi} \\ &= (1/6)\epsilon_{\rho\sigma\tau\theta}\, \epsilon_{\kappa\sigma\tau\theta} = \delta^{\rho}_{\kappa}, \end{aligned} \tag{D5}$$

and the $g^{\rho\mu}$ are therefore the contravariant components of the metric, as defined earlier.

One last preliminary result is not, strictly speaking, algebraic. We denote by D the partial derivative when the independent variables are the $g_{\sigma\tau}$ with $\sigma \leq \tau$. For any μ and ν og

$$Dg_{\mu\nu}/Dg_{\sigma\tau} = \zeta(\delta_{\mu\sigma}\delta_{\nu\tau} + \delta_{\mu\tau}\delta_{\nu\sigma}), \tag{D6}$$

where $\zeta = 1/2$ if $\sigma = \tau$, and $\zeta = 1$ if $\sigma < \tau$. We regard G as a function of the $g_{\sigma\tau}$ with $\sigma \leq \tau$, differentiate (D3), and find that

$$DG/Dg_{\sigma\tau} = 2\zeta G g^{\sigma\tau}. \tag{D7}$$

(Write out the four terms and use the skewsymmetry of $\epsilon_{\mu\nu\lambda\pi}$ to show that they are equal; then substitute from (D4).) Since G nowhere vanishes, (D7) implies that $D|G|/Dg_{\sigma\tau} = 2\zeta|G|g^{\sigma\tau}$. We define $\Delta = |G|^{1/2}$, so that $\Delta^2 = |G|$, and find that

$$D\Delta/Dg_{\sigma\tau} = \zeta\Delta g^{\sigma\tau}. \tag{D8}$$

If one multiplies (D8) by $g_{\sigma\tau,\pi}$, and notes that $g^{\sigma\tau}$ is symmetric, og

$$2\Delta_{,\pi} = \sum_{\sigma\leq\tau} 2(D\Delta/Dg_{\sigma\tau})g_{\sigma\tau,\pi} = \Delta g^{\sigma\tau} g_{\sigma\tau,\pi}. \tag{D9}$$

One can also prove (D9) by directly differentiating (D3) with respect to y^{π}.

Covariant divergence

We are now ready to prove that if a^μ are the components of a vector field, then the quantity $\Delta^{-1}(\Delta a^\mu)_{,\mu}$ is an invariant. We introduce the notation $Y_\nu^\mu = \partial y^\mu/\partial y'^\nu$, $Y_\nu'^\mu = \partial y'^\mu/\partial y^\nu$. The transformation laws for the vector field components a^μ and the partial derivatives ∂_μ are $a'^\mu = Y_\nu'^\mu a^\nu$ and $\partial'_\mu = Y_\mu^\nu \partial_\nu$; those of the metric components are $g'^{\rho\sigma} = Y_\pi'^\rho Y_\lambda'^\sigma g^{\pi\lambda}$ and $g'_{\rho\sigma} = Y_\rho^\pi Y_\sigma^\lambda g_{\pi\lambda}$. (Note that we are being sloppy here: we really should write $a'^\mu(y') = Y_\nu'^\mu(y)a^\nu(y)$, etc.). Eq.(D9) written for the chart y' gives

$$\Delta'^{-1}(\Delta' a'^\mu)_{,\mu} = \Delta'^{-1}\Delta'_{,\mu}a'^\mu + a'^\mu{}_{,\mu} = (1/2)g'^{\rho\sigma}g'_{\rho\sigma,\mu}a'^\mu + a'^\mu{}_{,\mu},$$

and noting that $Y_\nu'^\mu Y_\rho^\nu = Y_\nu^\mu Y_\rho'^\nu = \delta_\rho^\mu$, $g^{\rho\sigma}g_{\rho\tau} = g'^{\rho\sigma}g'_{\rho\tau} = \delta_\tau^\sigma$, and $Y'^\mu{}_{,\lambda\kappa} = Y'^\mu{}_{,\kappa\lambda}$ og

$$\Delta'^{-1}(\Delta' a'^\mu)_{,\mu}(y') = \Delta^{-1}(\Delta a^\mu)_{,\mu}(y). \tag{D10}$$

Exercise D1 Complete the derivation of (D10). ■

If $F^{\mu\nu}$ are the components of a tensor field and b_μ are those of a covector field, then $a^\mu = F^{\mu\nu}b_\nu$ are those of a vector field and (D10) shows that $\Delta^{-1}(\Delta F^{\mu\nu}b_\nu)_{,\mu} = \Delta^{-1}(\Delta F^{\mu\nu})_{,\mu}b_\nu + F^{\mu\nu}b_{\nu,\mu}$ is an invariant. The transformation law of the $F^{\mu\nu}$ is the same as that of the $g^{\mu\nu}$, and og $F'^{\mu\nu}b'_{\nu,\mu} = F^{\mu\nu}b_{\nu,\mu} + F^{\lambda\pi}Y^\rho_{\nu,\lambda}Y_\pi'^\nu b_\rho$, and hence

$$\Delta^{-1}(\Delta F^{\mu\rho})_{,\mu}b_\rho = \Delta'^{-1}(\Delta' F'^{\mu\nu})_{,\mu}Y_\nu^\rho b_\rho + F^{\lambda\pi}Y^\rho_{\nu,\lambda}Y_\pi'^\nu b_\rho. \tag{D11}$$

If $F^{\lambda\pi} = -F^{\pi\lambda}$, the last term vanishes: $F^{\lambda\pi}Y^\rho_{\nu,\lambda}Y_\pi'^\nu = -F^{\lambda\pi}Y_\nu^\rho Y'^\nu_{\pi,\lambda} = -F^{\lambda\pi}Y_\nu^\rho\partial^2 Y'^\nu/\partial y^\pi\partial y^\lambda = 0$. Since the b_ρ are arbitrary og

$$\begin{aligned}\Delta^{-1}(\Delta F^{\mu\rho})_{,\mu}(y) &= \Delta'^{-1}(\Delta' F'^{\mu\nu})_{,\mu}Y_\nu^\rho(y'),\\ \Delta'^{-1}(\Delta' F'^{\mu\pi})_{,\mu}(y') &= Y_\rho'^\pi\Delta^{-1}(\Delta F^{\mu\rho})_{,\mu}(y).\end{aligned} \tag{D12}$$

We have therefore proved that if F is a skewsymmetric tensor field, then $\Delta^{-1}(\Delta F^{\mu\rho})_{,\mu}$ are the components of a vector field. As an immediate corollary, the equation $(\Delta F^{\mu\rho})_{,\mu} = 0$ is covariant: if it holds in the chart y, it holds in any other chart y'. These results are important in electromagnetism, where F is the skewsymmetric, electromagnetic field tensor. Symmetric tensor fields are considered in the next exercise.

Exercise D2 Let $T^{\mu\nu}$ be the components of a symmetric tensor field, so that

$T^{\mu\nu} = T^{\nu\mu}$. Define $T^{\nu}_{\mu} = g_{\mu\pi}T^{\pi\nu}$, and show that

$$t_{\rho} = \Delta^{-1}(\Delta T^{\mu}_{\rho})_{,\mu} - (1/2)g_{\pi\sigma,\rho}T^{\pi\sigma} \tag{D13}$$

are the components of a covector field. [Use (D10) for the vector field with components $a^{\nu}T^{\mu}_{\nu}$.] ■

The expressions that we have derived are usually called *covariant divergences*, and are written as

$$\begin{aligned} a^{\mu}{}_{;\mu} &= \Delta^{-1}(\Delta a^{\mu})_{,\mu}, \\ F^{\mu\rho}{}_{;\mu} &= \Delta^{-1}(\Delta F^{\mu\rho})_{,\mu}, \\ T^{\mu}_{\rho;\mu} &= \Delta^{-1}(\Delta T^{\mu}_{\rho})_{,\mu} - (1/2)g_{\pi\sigma,\rho}T^{\pi\sigma}. \end{aligned} \tag{D14}$$

Eq.(D10), for example, is then equivalent to $a'^{\mu}{}_{;\mu}(y') = a^{\mu}{}_{;\mu}(y)$, or to the statement that $a^{\mu}{}_{;\mu}$ is an invariant function. Similarly, $F^{\mu\rho}{}_{;\mu}$ are the components of a contravariant vector field, and $T^{\mu}_{\rho;\mu}$ are the components of a covariant vector field.

Covariant derivative

The partial derivatives of the components of a tensor field are not themselves the components of a tensor field, in general. However, for the special case of inertial charts in special relativity, they *are* components of a tensor field. For example, if a^{μ} and a'^{μ} are the components of a vector field a in the inertial charts x and x', respectively, then $M^{\mu}_{\nu} = \partial a^{\mu}/\partial x^{\nu}$ and $M'^{\mu}_{\nu} = \partial a'^{\mu}/\partial x'^{\nu}$ are the components of a tensor field M in x and x'. We are going to define a generalization of the partial derivative, called the *covariant derivative*. The covariant derivative of a tensor field is always a tensor field, for any metric and any chart. In special relativity, the covariant derivative reduces to the partial derivative in an inertial chart.

In any chart y, the *Christoffel symbols* of the first and second kinds are defined by

$$[\mu\nu, \pi] = (1/2)(g_{\mu\pi,\nu} + g_{\nu\pi,\mu} - g_{\mu\nu,\pi}), \qquad \left\{{\rho \atop \mu\nu}\right\} = g^{\rho\pi}[\mu\nu, \pi]. \tag{D15}$$

Note that they are symmetric in the indices μ and ν. This implies that $\left\{{\sigma \atop \sigma\pi}\right\} = (1/2)g^{\sigma\tau}g_{\sigma\tau,\pi}$, and (D9) can be written as

$$\Delta_{,\pi} = \left\{{\sigma \atop \sigma\pi}\right\}\Delta. \tag{D16}$$

If χ is an invariant, the partial derivatives $\chi_{,\mu}$ are the components of a covector field. We define this covector field to be the covariant derivative of χ, and write $\chi_{,\mu} = \chi_{;\mu}$.

The covariant derivative of the vector field with components a^μ is the tensor field with components $a^\mu{}_{;\nu}$ given by

$$a^\mu{}_{;\nu} = a^\mu{}_{,\nu} + a^\kappa \{{}^{\ \mu}_{\kappa\nu}\}. \tag{D17}$$

The covariant derivative of the covector field with components b_μ is the tensor field with components $b_{\mu;\nu}$ given by

$$b_{\mu;\nu} = b_{\mu,\nu} - b_\kappa \{{}^{\ \kappa}_{\mu\nu}\}. \tag{D18}$$

The covariant derivatives of tensor fields of higher order are found in a similar manner. Each contravariant (superscript) index gives a Christoffel symbol term like that in (D17); each covariant (subscript) index gives one like that in (D18). For example, og

$$T^{\mu\nu}{}_{;\pi} = T^{\mu\nu}{}_{,\pi} + T^{\kappa\nu}\{{}^{\ \mu}_{\kappa\pi}\} + T^{\mu\kappa}\{{}^{\ \nu}_{\kappa\pi}\}, \tag{D19}$$

$$T^\mu_{\nu;\pi} = T^\mu_{\nu,\pi} + T^\kappa_\nu\{{}^{\ \mu}_{\kappa\pi}\} - T^\mu_\kappa\{{}^{\ \kappa}_{\nu\pi}\}. \tag{D20}$$

The covariant derivative of a tensor field has always the same contravariant order as the original tensor field, and covariant order greater by 1. (The contravariant order is the number of superscripts in a component; the covariant order is the number of subscripts.)

Exercise D3 Derive (D18) from (D17) and the assumption that

$$(a^\mu b_\mu)_{;\nu} = a^\mu{}_{;\nu} b_\mu + a^\mu b_{\mu;\nu}. \ \blacksquare \tag{D21}$$

To prove that (D17), (D18), etc., do transform like the components of tensor fields, one should first derive the transformation laws for the Christoffel symbols.

Exercise D4 Define $Y^\rho_{\nu,\sigma} = (\partial/\partial y^\sigma)Y^\rho_\nu$, and show that

$$[\mu\nu,\pi]' = Y^\rho_\mu Y^\sigma_\nu Y^\tau_\pi [\rho\sigma,\tau] + Y^\rho_{\nu,\sigma} Y^\sigma_\mu Y^\tau_\pi g_{\rho\tau}, \tag{D22}$$

$$\{{}^{\ \kappa}_{\mu\nu}\}' = g'^{\kappa\pi}[\mu\nu,\pi]' = Y'^\kappa_\iota Y^\rho_\mu Y^\sigma_\nu \{{}^{\ \iota}_{\rho\sigma}\} + Y'^\kappa_\iota Y^\sigma_\mu Y^\iota_{\nu,\sigma}. \tag{D23}$$

[Note that $Y^\rho_{\nu,\sigma} Y^\sigma_\mu = Y^\rho_{\mu,\sigma} Y^\sigma_\nu = \partial^2 y^\rho / \partial y'^\mu \partial y'^\nu$. The proof is in [Eisenhart 26], P.19: old books like this often contain proofs that are omitted in more modern works.] ■

The properties of the covariant derivative are very similar to those of the partial derivative. We illustrate them by examples. For any invariant function f and any 2 contravariant tensors S and T, og

$$(fT^{\mu\nu} + S^{\mu\nu})_{;\pi} = f_{,\pi}T^{\mu\nu} + fT^{\mu\nu}{}_{;\pi} + S^{\mu\nu}{}_{;\pi}, \tag{D24}$$

$$(T^{\mu\nu}S^{\rho\sigma})_{;\pi} = T^{\mu\nu}{}_{;\pi}S^{\rho\sigma} + T^{\mu\nu}S^{\rho\sigma}{}_{;\pi}. \tag{D25}$$

One easily proves that the covariant derivative of the metric tensor vanishes:

$$g^{\mu\nu}{}_{;\pi} = 0, \qquad g_{\mu\nu;\pi} = 0. \tag{D26}$$

If b_μ and $T^{\mu\nu}$ are tensor-field components, then $a^\nu = b_\mu T^{\mu\nu}$ are vector field components (we say that a is the *contraction* of b and T). One shows that

$$a^\nu{}_{;\pi} = (b_\mu T^{\mu\nu})_{;\pi} = b_{\mu;\pi}T^{\mu\nu} + b_\mu T^{\mu\nu}{}_{;\pi} \tag{D27}$$

(i.e. the covariant derivative commutes with the operation of contraction). If b_μ are the components of a covector field, then $b^\mu = g^{\mu\nu}b_\nu$ are the components of a vector field, and from (D26) and (D27) og

$$b^\mu{}_{;\pi} = (g^{\mu\nu}b_\nu)_{;\pi} = g^{\mu\nu}{}_{;\pi}b_\nu + g^{\mu\nu}b_{\nu;\pi} = g^{\mu\nu}b_{\nu;\pi}. \tag{D28}$$

Similarly, one shows that $b_{\mu;\pi} = g_{\mu\nu}b^\nu{}_{;\pi}$.

Armed with a little faith, the reader should now be able to follow even Chapter IX. She would be well advised, however, to study some differential geometry. She will find that the clumsy calculations that we have outlined here can be replaced by elegant axiomatics.

E Relations between Centres of Mass

In Chapter VIII, we defined centres of mass in terms of the conserved mass density and the inertial mass density. We now consider a mass density that differs by an arbitrary, small amount from the conserved mass density, and calculate the acceleration of the corresponding centre of mass.

We again denote the conserved mass density by ρ^*. We define a mass density $\rho' = \rho^*(1+\alpha)$, where α is any smooth, O_2 function of the spacetime variables. The mass M'_B of body B is defined by

$$M'_B = \{\rho'\}_B = M^*_B + \{\rho^*\alpha\}_B. \tag{E1}$$

The centre of mass corresponding to ρ^* is defined by VIII(1.1): $M^*_B X^m_B = \{\rho^* x^m\}_B$; that corresponding to ρ' is defined similarly by

$$M'_B X'^m_B = \{\rho' x^m\}_B = M^*_B X^m_B + \{\rho^*\alpha x^m\}_B. \tag{E2}$$

The difference between the two centres of mass is given by

$$\begin{aligned} M^*_B(X'^m_B - X^m_B) &= (M^*_B - M'_B)X'^m_B + M'_B X'^m_B - M^*_B X^m_B \\ &= \{\rho^*\alpha(x^m - X'^m_B)\}_B. \end{aligned} \tag{E3}$$

We define the velocity and acceleration of the ρ' centre of mass of B by $\boldsymbol{V}'_B = D\boldsymbol{X}'_B$ and $\boldsymbol{A}'_B = D\boldsymbol{V}'_B$, respectively. Differentiating (E3) and using VII(4.13), og

$$\begin{aligned} M^*_B(V'^m_B - V^m_B) &= D\big\{\rho^*\alpha(x^m - X'^m_B)\big\}_B \\ &= \big\{\rho^*(D_t\alpha)(x^m - X'^m_B)\big\}_B + \big\{\rho^*\alpha(V^m - V'^m_B)\big\}_B, \end{aligned} \tag{E4}$$

$$\begin{aligned} M^*_B(A'^m_B - A^m_B) = &\big\{\rho^*(D_t^2\alpha)(x^m - X'^m_B)\big\}_B \\ &+ 2\big\{\rho^*(D_t\alpha)(V^m - V'^m_B)\big\}_B + \big\{\rho^*\alpha(A^m - A'^m_B)\big\}_B. \end{aligned} \tag{E5}$$

Since $M_B^*(A_B'^m - A_B^m) = M_B' A_B'^m - M_B^* A_B^m + (M_B^* - M_B')A_B'^m = M_B' A_B'^m - M_B^* A_B^m - \{\rho^*\alpha\}_B A_B'^m$, og

$$M_B' A_B'^m = M_B^* A_B^m + \left\{\rho^*(D_t^2\alpha)(x^m - X_B'^m)\right\}_B + 2\left\{\rho^*(D_t\alpha)(V^m - V_B'^m)\right\}_B + \left\{\rho^*\alpha A^m\right\}_B. \tag{E6}$$

One can replace $X_B'^m$ by X_B^m and $V_B'^m$ by V_B^m in (E6) with an error of order $\rho c^2 L^2 O_4$. Defining $\Delta_B^m(\alpha) = M_B' A_B'^m - M_B^* A_B^m$, og

$$\Delta_B^m(\alpha) = \left\{\rho^*(D_t^2\alpha)(x^m - X_B^m)\right\}_B + 2\left\{\rho^*(D_t\alpha)(V^m - V_B^m)\right\}_B + \left\{\rho^*\alpha A^m\right\}_B + \rho c^2 L^2 O_4. \tag{E7}$$

Eq.(E7) allows us to calculate $\Delta_B^m(\alpha)$ for any choice of α. Note that Δ_B^m is linear:

$$\Delta_B^m(\alpha + \lambda\beta) = \Delta_B^m(\alpha) + \lambda\Delta_B^m(\beta) \tag{E8}$$

for any functions α and β and any real number λ.

Exercise E1 Calculate $\Delta_B^m(\alpha)$ for $\alpha = \Pi$, Υ, $\boldsymbol{V}\cdot\boldsymbol{V}_B$, U_B (see Chapter VII and VIII(2.3), VIII(4.11)), and also for $\alpha = f$, where f is a function of t only. ■

Exercise E2 Use the results of the last exercise to calculate $\Delta_B^m(\alpha)$ for $\alpha = (1/2)c^{-2}|\boldsymbol{V} - \boldsymbol{V}_B|^2 - (1/2)U_B + \Pi$. [This is the α that corresponds to $M_B' = m_B$, where m_B is the inertial mass VIII(4.12).] ■

Exercise E3 Use the results of Exercise E1 and Exercise 4.1 of Chapter VIII to calculate $\boldsymbol{A}_B$, the acceleration of the centre of mass defined in terms of the conserved mass density. ■

Exercise E4 Calculate $\Delta_B^m(\alpha)$ for $\alpha = 3\Upsilon + 3p\rho^{-1}c^{-2} - U + \Pi$ and compare the acceleration of the centre of mass with the previous results. [This is the α that corresponds to the gravitational mass $\mathcal{M}_B$ – see V(2.6) and VIII(4.37).] ■

F Perturbation Theory for the Kepler Problem

The idea of perturbation theory is to start from a system whose motion under given forces is known, to change the forces by a small amount, and to calculate approximately the change in the motion. The original system is called the *unperturbed system*, the additional forces are called the *perturbation*, and the system with the additional forces is the *perturbed system*. We are going to consider the very special case of the Kepler problem: the unperturbed system consists of two small, or spherically symmetric, bodies moving under their mutual, Newtonian, gravitational attraction.

We start from the solution of the equation of motion $D^2\boldsymbol{R} = -KR^{-3}\boldsymbol{R}$, where $\boldsymbol{R}$ is the relative displacement of two bodies, $R = |\boldsymbol{R}|$, $K > 0$ is a constant, and $D = d/dt$, as in Chapter VIII. The quantities $E, \boldsymbol{L}$, and $\boldsymbol{e}$, defined by VIII(3.7), are all constant for the unperturbed motion, and VIII(3.8)–VIII(3.10) are satisfied.

The perturbed equation of motion is

$$D^2\boldsymbol{R} = -KR^{-3}\boldsymbol{R} + \boldsymbol{f}, \tag{F1}$$

where the perturbation $\boldsymbol{f}$ has the dimensions of force per unit mass. We assume that $\boldsymbol{f}$ is small, in the sense that $fR^2/K \ll 1$, and we neglect terms quadratic in the components of $\boldsymbol{f}$ – this is called *first-order perturbation theory*.

We remark that second-order, and higher-order, perturbation theory has been developed, and was a preoccupation of astronomers in the nineteenth century. It is sometimes unreliable, and has been largely supplanted by numerical integration techniques [Taff 85]. However, the first-order theory is usually trouble-free and often sufficiently accurate.

When $\boldsymbol{f} \neq 0$, the quantities $E, \boldsymbol{L}$, and $\boldsymbol{e}$ may be functions of time. From (F1),

og

$$\begin{aligned} DE = \boldsymbol{v}\cdot\boldsymbol{f}, \qquad D\boldsymbol{L} = \boldsymbol{R}\times\boldsymbol{f}, \\ KD\boldsymbol{e} = -(\boldsymbol{R}\cdot\boldsymbol{v})\boldsymbol{f} + 2(\boldsymbol{f}\cdot\boldsymbol{v})\boldsymbol{R} - (\boldsymbol{R}\cdot\boldsymbol{f})\boldsymbol{v}. \end{aligned} \tag{F2}$$

For any vector $\boldsymbol{x}$, we define $x = |\boldsymbol{x}|$, and the unit vector $\hat{x} = \boldsymbol{x}x^{-1}$. We assume that $L \neq 0$ and $0 < e < 1$, and define $\hat{\jmath} = \hat{L} \times \hat{e}$, so that $(\hat{e}, \hat{\jmath}, \hat{L})$ is a right-handed, orthonormal basis. We choose cylindrical polar coordinates R, θ, z such that $\boldsymbol{R} = \hat{e}R\cos\theta + \hat{\jmath}\sin\theta + \hat{L}z$. When $\boldsymbol{f} = 0$, one shows that $L = R^2 D\theta$ and $R(1 + e\cos\theta) = L^2/K = \lambda$, and that z is constant.

We regard the orbit as a curve that is parameterized by θ. We suppose that g is a given function which is defined on the orbit for $\theta \in [0, 2\pi]$. (In practice, g is often given as a function of $\boldsymbol{R}, \boldsymbol{v}$, etc., which are themselves functions of θ on the orbit). We define Δg, the *change in g per orbit*, by

$$\Delta g = \int_0^{2\pi} (dg/d\theta)\, d\theta. \tag{F3}$$

Let g be a constant of the unperturbed motion. One can then calculate Δg for the perturbed motion, to first order in $\boldsymbol{f}$, by evaluating $dg/d\theta$ on the *unperturbed orbit*. We write $\boldsymbol{f} = f_1\hat{e} + f_2\hat{\jmath} + f_3\hat{L}$, and show after some short calculations that

$$\Delta E = L^2K^{-1} \int_0^{2\pi} \Big[-f_1\sin\theta + f_2(e + \cos\theta)\Big](1 + e\cos\theta)^{-2}\, d\theta, \tag{F4}$$

$$\Delta \boldsymbol{L} = L^5K^{-3} \int_0^{2\pi} \Big[(f_2\cos\theta - f_1\sin\theta)\hat{L} + f_3\sin\theta\hat{e} - f_3\cos\theta\hat{\jmath}\Big]$$

$$(1 + e\cos\theta)^{-3} d\theta, \tag{F5}$$

$$\Delta \boldsymbol{e} = L^4K^{-3} \int_0^{2\pi} \Big\{-f_3 e\sin\theta\hat{L}$$

$$+ \Big[-f_1\sin\theta(e + \cos\theta) + f_2(1 + \cos^2\theta + 2e\cos\theta)\Big]\hat{e}$$

$$+ \Big[f_2\sin\theta\cos\theta - f_1(1 + \sin^2\theta + e\cos\theta)\Big]\hat{\jmath}\Big\}(1 + e\cos\theta)^{-3} d\theta, \tag{F6}$$

where e and L have their unperturbed values.

Since $\boldsymbol{e} \cdot D\boldsymbol{e} = eDe$, og $\Delta e = \hat{e} \cdot \Delta\boldsymbol{e}$, and similarly $\Delta L = \hat{L} \cdot \Delta\boldsymbol{L}$. From $D\boldsymbol{e} = \hat{e}De + eD\hat{e}$, og $\Delta\boldsymbol{e} = \hat{e}\Delta e + e\Delta\hat{e}$, and hence $\hat{\jmath} \cdot \Delta\boldsymbol{e} = e\hat{\jmath} \cdot \Delta\hat{e}$. The advance in perihelion per orbit is $\hat{\jmath} \cdot \Delta\boldsymbol{e}$.

Exercise F1 Calculate ΔE, $\Delta\boldsymbol{L}$, and $\Delta\boldsymbol{e}$ for the radial perturbation $\boldsymbol{f}(\boldsymbol{R}) = h(R)\hat{R}$, where h is a smooth function of R. Hence find the perturbation of a planetary orbit due to the rotation of the Sun. [Assume that the axis of rotation of the Sun is normal to the plane of the planet's orbit (this is not quite true!). Use the expression for the quadrupole gravitational field given in Chapter VIII, Section 1.] ■

In order to calculate the quadrupole moment of the Sun, one must know how it rotates. There was evidence, based on measurements of the shape of the solar disc, that the Sun's core rotates much more rapidly than the surface. However, more recent measurements of the Sun's oscillations indicate that this is not so. The case of rotation with a small, constant angular velocity is the subject of the next exercise.

Exercise F2 A ball of ideal fluid of mass M and constant density ρ rotates with constant angular velocity $\boldsymbol{\omega}$ in a Galilean frame. Calculate the shape of the ball and its gravitational quadrupole moment. [It is best to work in the rotating frame where the fluid is at rest. The total potential for the forces acting on the fluid is the sum of the gravitational potential and the centrifugal potential $-(1/2)|\boldsymbol{\omega}|^2 z^2$, where z is the perpendicular distance from the field point to the rotation axis. The total potential (gravitational plus centrifugal) is constant on the surface of the fluid. For small $|\boldsymbol{\omega}|$, one may assume that the cross-sections of the ball through the axis of rotation are ellipses with semi-axes a and $a(1+\epsilon)$. Neglecting terms in ϵ^2, one easily calculates the quadrupole moment.] ■

G Langrangian Field Theory

Here are proofs of the results used in Chapter IX: the Euler-Lagrange equations and the vanishing of the covariant divergence of the stress-momentum. To simplify matters, we ignore problems concerning the domains of spacetime charts. In general, a spacetime chart Y is not defined at all spacetime points but only in a restricted region, called its *domain*. Another chart Y' may have a different domain, and one can define transformations from Y to Y' only for points that lie in both domains. The proper, mathematical context for such questions is the theory of differential manifolds, which we shall not discuss. We simply assume that all domains are sufficiently large and all transformations sufficiently smooth.

Unless otherwise stated, lower-case Greek indices have the range $\{0, 1, 2, 3\}$ and upper-case Latin the range $\{1, 2, \ldots, M\}$. Both obey the summation convention.

Let Y be a spacetime chart with coordinate functions y^μ , and u_A be real functions – the *fields* or *field components*. The u_A are C^2 functions of $y = (y^0, y^1, y^2, y^3)$; their partial derivatives with respect to y^μ are denoted by $u_{A,\mu}$. The components of the metric in Y are $g_{\mu\nu}$, and we define $\Delta = |\det g_{\mu\nu}|^{1/2}$. Note that we have refined our notation. Previously, we often used the same symbol y for the spacetime chart and the quadruple (y^0, y^1, y^2, y^3). This can lead to ambiguity: in (G1), for example, it would be misleading to write $J(u, y, \Omega)$.

We assume that there is an invariant, C^2 function L of the fields u_A, their first partial derivatives $u_{A,\mu}$, and y, and we define a functional J on a region Ω of spacetime by

$$J(u, Y, \Omega) = \int_{y(\Omega)} L(u(y), Du(y), y)\Delta\, d^4y, \tag{G1}$$

where $u = (u_1, \ldots, u_M), Du = (u_{1,0}, u_{1,1}, \ldots, u_{M,2}, u_{M,3})$, $d^4y = dy^0dy^1dy^2dy^3$. The region Ω must be chosen so that the integral exists; we define $y(\Omega)$ to be the set of all quadruples $y = (y^0, y^1, y^2, y^3)$ that correspond to spacetime points in Ω.

What does it mean to say that L is an *invariant function*? Let us suppose that Y' is any chart that is smoothly related to Y (that is, the y'^{μ} are C^2 functions of the y^{ν}, and the y^{μ} are C^2 functions of the y'^{ν}). We write the components of the metric in Y' as $g'_{\mu\nu}$, and the field components in Y' as u'_A, and we define $u'_{A,\mu}(y') = (\partial/\partial y'^{\mu})u'_A(y')$, $y' = (y'^0, y'^1, y'^2, y'^3)$, $u' = (u'_1, \ldots, u'_M)$, $D'u = (u'_{1,0}, u'_{1,1}, \ldots, u'_{M,2}, u'_{M,3})$. We then define L to be an invariant function if $L(u(y), Du(y), y) = L(u'(y'), D'u'(y'), y')$.

In order to construct an invariant function, it is often necessary to include the $g_{\mu\nu}$ among the field components u_A. More generally, the $g_{\mu\nu}$ may be functions of the u_A and $u_{A,\pi}$. We need not include the $g_{\mu\nu}$ among the field components if we can restrict ourselves to charts in which the $g_{\mu\nu}$ have always the same, constant values – the inertial charts of special relativity, for example.

The components of the metric in Y and Y' are related by $g'_{\mu\nu}(y') = (\partial y^{\pi}/\partial y'^{\mu})$. $(\partial y^{\rho}/\partial y'^{\nu})g_{\pi\rho}(y)$. It follows that $\det g'_{\mu\nu} = \Gamma^2 \det(g_{\mu\nu})$, where $\Gamma = \det(\partial y^{\pi}/\partial y'^{\mu})$ is the Jacobian, and that $\Delta' = |\det g'_{\mu\nu}|^{1/2} = |\Gamma|\Delta$. By the usual rule for transforming multiple integrals, og

$$\begin{aligned} J(u', Y', \Omega) &= \int_{y'(\Omega)} L(u'(y'), D'u'(y'), y')\Delta'\, d^4y' \\ &= \int_{y(\Omega)} L(u(y), Du(y), y)\Delta\, d^4y = J(u, Y, \Omega), \end{aligned} \qquad \text{(G2)}$$

and we say that J is an *invariant functional*. We say that J is a *functional* rather than a *function* because, unlike L, it does not have a local dependence on u: we cannot write $J(u, Y, \Omega)(y) = J(u(y), Y, \Omega)$.

We assume that the fields $\hat{u}$ that describe a possible physical system are such that $J(\hat{u}, y, \Omega)$ is an extremum for any acceptable Ω. To make this more precise, let $h(y) = (h_1(y), \ldots, h_M(y))$, where the h_A are C^2 functions, and let $h(y) = 0$ on the boundary of Ω. Define $j(\lambda) = J(\hat{u} + \lambda h, Y, \Omega)$ for all real λ in some interval that contains zero. Then J is an *extremum* (at $\hat{u}$), and $\hat{u}$ is an *extremal* of J, if $Dj(0) = 0$ for all h and Ω.

To simplify the notation, we define the *Lagrangian density* $\mathcal{L}$ by $\mathcal{L} = L\Delta$, as in Chapter IX. Since Δ is a function of the the $g_{\mu\nu}$, we can write it as $\Delta(u, Du)$ (or as $\Delta(u)$ if the $g_{\mu\nu}$ depend only on the u_A). Hence $\mathcal{L}$ is a function of u, Du, and y, and the condition for J to be an extremum at $\hat{u}$ can be written as

$$(d/d\lambda)_{\lambda=0} \int_{y(\Omega)} \mathcal{L}(\hat{u}(y) + \lambda h(y), D\hat{u}(y) + \lambda Dh(y), y)\, d^4y = 0. \qquad \text{(G3)}$$

Note that Ω is taken to be independent of λ. Under our assumptions, one can reverse the order of integration and differentiation with respect to λ, and one finds that the condition for an extremum is

$$\int_{y(\Omega)} \left[h_A(y)\hat{\mathcal{L}}_{u_A} + h_{A,\mu}(y)\hat{\mathcal{L}}_{u_{A,\mu}}\right] d^4y = 0, \tag{G4}$$

where $\hat{\mathcal{L}}_{u_A} = (\partial/\partial u_A)\mathcal{L}(\hat{u}(y), D\hat{u}(y), y)$, $\hat{\mathcal{L}}_{u_{A,\mu}} = (\partial/\partial u_{A,\mu})\mathcal{L}(\hat{u}(y), D\hat{u}(y), y)$. Writing $h_{A,\mu}(y)\hat{\mathcal{L}}_{u_{A,\mu}} = (h_A(y)\hat{\mathcal{L}}_{u_{A,\mu}})_{,\mu} - h_A(y)(\hat{\mathcal{L}}_{u_{A,\mu}})_{,\mu}$, and recalling that $h_A(y)$ vanishes on the boundary of Ω, og

$$\int_{y(\Omega)} \left[h_A(y)\hat{\mathcal{L}}_{u_{A,\mu}}\right]_{,\mu} d^4y = 0 \tag{G5}$$

by the divergence theorem, and (G4) becomes

$$\int_{y(\Omega)} h_A(y)\left[\hat{\mathcal{L}}_{u_A} - (\hat{\mathcal{L}}_{u_{A,\mu}})_{,\mu}\right] d^4y = 0. \tag{G6}$$

Since the bracket is continuous as a function of the y^μ, and the h_A are arbitrary, it follows that

$$\hat{\mathcal{L}}_{u_A} - (\hat{\mathcal{L}}_{u_{A,\mu}})_{,\mu} = 0. \tag{G7}$$

These are the *Euler-Lagrange equations*. They are a necessary condition for J to be an extremum at û. One often speaks of (G7) as the *field equations* (for the u_A).

The conditions that we have derived are sometimes too stringent, in the sense that there are no fields $\hat{u}_A$ for which they are satisfied. In special relativity, for example, the components of the metric in the chart Y are uniquely determined functions, quite independent of any physical fields that may be present in spacetime. If one tries to find this metric from an extremum of J, one must expect to get nonsense. In many theories of gravitation, on the other hand, the components of the metric determine the gravitational field, and one can hope to derive field equations for them by a variational procedure.

As in Chapter IX, we distinguish between *dynamical fields*, which describe particular physical systems, and *absolute fields* such as the components of the metric in special relativity, which describe the spacetime in which physical fields evolve, but which are independent of the particular physical fields that happen to be present. Mathematically speaking, the distinction is between the u_A that are subject to variation, and those that are not. Let us change our previous notation, and make the convention that u_A for $A \in \{1, \ldots, M\}$ are dynamical fields, and u_A for $A \in \{M+1, \ldots, M+M'\}$ are absolute fields. The Euler-Lagrange equations are then to be satisfied only for $A \in \{1, \ldots, M\}$.

In what follows, we usually drop the caret from the field components in the Euler-Lagrange equations. One must understand, however, that these equations hold only on the extremals (the caret is still there, although invisible!).

The stress-momentum

We are going to prove that the Euler-Lagrange equations and the invariance of the Lagrangian imply, under certain conditions, the vanishing of the covariant divergence of the stress-momentum. To do this, we consider a class of spacetime charts that are "close to" a given chart Y. More precisely, let $Y(\tau)$ be a chart for all real τ in some interval that includes zero, and let $Y(0) = Y$; the coordinates of Y are y and those of $Y(\tau)$ are y'. We assume that $y' = y + \tau w(y)$, where $w(y) = (w^0(y), w^1(y), w^2(y), w^3(y))$, and the w^μ are C^2 functions.

Since $\partial y'^\pi/\partial y^\mu = \delta^\pi_\mu + \tau w^\pi_{,\mu}(y)$, og $\partial y^\pi/\partial y'^\mu = \delta^\pi_\mu - \tau w^\pi_{,\mu}(y) + O(\tau^2)$ as $\tau \to 0$, and $\Gamma = \det(\partial y^\pi/\partial y'^\mu) = 1 - \tau w^\pi_{,\pi}(y) + O(\tau^2)$. If $g'_{\mu\nu}$ are the components of the metric in y', og

$$\Delta' = |\det(g'_{\mu\nu})|^{1/2} = |\Gamma|\Delta = (1 - \tau w^\pi_{,\pi})\Delta + O(\tau^2), \tag{G8}$$

as $\tau \to 0$.

The components of the metric in Y and $Y(\tau)$ are related by

$$\begin{aligned} g'_{\mu\nu}(y') &= (\partial y^\pi/\partial y'^\mu)(\partial y^\rho/\partial y'^\nu) g_{\pi\rho}(y). \\ &= (\delta^\pi_\mu - \tau w^\pi_{,\mu}(y))(\delta^\rho_\nu - \tau w^\rho_{,\nu}(y)) g_{\pi\rho}(y) + O(\tau^2) \\ &= g_{\mu\nu}(y) - \tau(w^\pi_{,\mu} g_{\pi\nu} + w^\pi_{,\nu} g_{\pi\mu})(y) + O(\tau^2) \end{aligned} \tag{G9}$$

as $\tau \to 0$. The field components in $Y(\tau)$ are written as u'_A. We assume that there are functions B^B_A, where $A, B \in \{1, \ldots, M + M'\}$, such that

$$u'_A(y') = u_A(y) + \tau B^B_A(y) u_B(y) + 0(\tau^2) \tag{G10}$$

as $\tau \to 0$, and that the B^B_A are linear functions of the $w^\rho_{,\pi}$ (you should check that this is true if the u_A are the components of a tensor field). Differentiating with respect to y'^μ, og

$$\begin{aligned} u'_{A,\pi}(y') &= u_{A,\pi}(y) - \tau u_{A,\rho} w^\rho_{,\pi}(y) \\ &\quad + \tau B^B_{A,\pi}(y) u_B(y) + \tau B^B_A(y) u_{B,\pi}(y) + 0(\tau^2). \end{aligned} \tag{G11}$$

For brevity, we define $\delta u_A = u'_A(y') - u_A(y)$ and $\delta u_{A,\pi} = u'_{A,\pi}(y') - u_{A,\pi}(y)$, and similarly for $\delta g_{\mu\nu}$ and $\delta g_{\mu\nu,\pi}$. Trivially og

$$\begin{aligned} \delta u_{A,\pi} &= (\delta u_A)_{,\pi} - \tau u_{A,\rho} w^{\rho}_{,\pi}(y) + 0(\tau^2), \\ \delta g_{\mu\nu,\pi} &= (\delta g_{\mu\nu})_{,\pi} - \tau g_{\mu\nu,\rho} w^{\rho}_{,\pi}(y) + 0(\tau^2), \end{aligned} \tag{G12}$$

where δu_A and $\delta g_{\mu\nu}$ are regarded as functions of y, and the comma denotes the partial derivative with respect to the y^{π}.

The assumed invariance of the Lagrangian L implies that

$$\delta L = L(u'(y'), D'u'(y'), y') - L(u(y), Du(y), y) = 0. \tag{G13}$$

Since $\mathcal{L} = \Delta L$, it follows that $\delta\mathcal{L} = \mathcal{L}(u'(y'), D'u'(y'), y') - \mathcal{L}(u(y),\ Du(y), y) = L\delta\Delta$, and hence from (G8)

$$\delta\mathcal{L} = -\mathcal{L}\tau w^{\pi}_{,\pi} + O(\tau^2) \tag{G14}$$

as $\tau \to 0$.

Using a similar notation for partial derivatives as in (G4), og

$$\delta\mathcal{L} = \mathcal{L}_{u_A}\delta u_A + \mathcal{L}_{u_{A,\mu}}\delta u_{A,\mu} + \mathcal{L}_{y^\mu}\delta y^{\mu}_{,} \tag{G15}$$

where δu_A and $\delta u_{A,\mu}$ are defined as above, and $\delta y^{\mu} = w^{\mu}\tau$. We substitute (G15) into (G14), note that

$$(\mathcal{L}w^{\pi})_{,\pi} = (\mathcal{L}_{u_A} u_{A,\pi} + \mathcal{L}_{u_{A,\mu}} u_{A,\mu\pi} + \mathcal{L}_{y^{\pi}})w^{\pi} + \mathcal{L}w^{\pi}_{,\pi}, \tag{G16}$$

and get

$$\mathcal{L}_{u_A}\delta u_A + \mathcal{L}_{u_{A,\mu}}\delta u_{A,\mu} - (\mathcal{L}_{u_A} u_{A,\pi} + \mathcal{L}_{u_{A,\mu}} u_{A,\mu\pi})w^{\pi}\tau + (\mathcal{L}w^{\pi})_{,\pi}\tau = O(\tau^2). \tag{G17}$$

Substituting from (G12), og

$$\mathcal{L}_{u_A}(\delta u_A - u_{A,\pi}w^{\pi}\tau) + \mathcal{L}_{u_{A,\mu}}(\delta u_A - u_{A,\pi}w^{\pi}\tau)_{,\mu} + (\mathcal{L}w^{\pi})_{,\pi}\tau = O(\tau^2), \tag{G18}$$

which is equivalent to

$$\left[\mathcal{L}_{u_A} - (\mathcal{L}_{u_{A,\mu}})_{,\mu}\right](\delta u_A - u_{A,\pi}w^{\pi}\tau) + \left[\mathcal{L}_{u_{A,\mu}}(\delta u_A - u_{A,\pi}w^{\pi}\tau)\right]_{,\mu} + (\mathcal{L}w^{\pi})_{,\pi}\tau = O(\tau^2). \tag{G19}$$

The index A in (G15)–(G19) is summed from 1 to $M + M'$; the u_A are dynamical variables for $A \in \{1, \ldots, M\}$ and absolute variables for $A \in \{M+1, \ldots, M+M'\}$. Note that we have not so far assumed that u is an extremal, and have not used the Euler-Lagrange equations. Note also that $\delta u_A - u_{A,\pi}w^{\pi}\tau = u'_A(y) - u_A(y) + O(\tau^2)$: this is the change in u_A when one goes from the point with coordinates y in Y to the point with the same coordinates in $Y(\tau)$.

For any region Ω, one can choose the w^π and $w^\pi_{,\rho}$ to vanish on the boundary of Ω. One then has $\delta u_A = 0$ on the boundary, from (G10). Integrating (G19) over $y(\Omega)$ and using Gauss' theorem, og

$$\int_{y(\Omega)} \Big[\mathcal{L}_{u_A} - (\mathcal{L}_{u_{A,\mu}})_{,\mu}\Big](\delta u_A - u_{A,\pi} w^\pi \tau)\, d^4y = O(\tau^2). \tag{G20}$$

On an extremal, (G7) holds for the dynamical variables, $A \in \{1, \ldots, M\}$, and (G20) gives

$$\sum_A \int_{y(\Omega)} \Big[\hat{\mathcal{L}}_{u_A} - (\hat{\mathcal{L}}_{u_{A,\mu}})_{,\mu}\Big](\delta \hat{u}_A - \hat{u}_{A,\pi} w^\pi \tau)\, d^4y = O(\tau^2), \tag{G21}$$

where the sum is for $A \in \{M+1, \ldots, M+M'\}$.

Our results so far are quite general. We now consider the special case when the $g_{\mu\nu}$ are absolute variables and there are no other absolute variables. We define a stress-momentum Θ by

$$\zeta \Delta \Theta^{\mu\nu} = \hat{\mathcal{L}}_{g_{\mu\nu}} - (\hat{\mathcal{L}}_{g_{\mu\nu,\pi}})_{,\pi} \tag{G22}$$

where $\zeta = 1/2$ if $\mu = \nu$ and $\zeta = 1$ if $\mu \neq \nu$ (cf. IX(3.9) and IX(3.10); in the notation of Chapter IX, we would write $D\hat{\mathcal{L}}/Dg_{\mu\nu} - (D\hat{\mathcal{L}}/Dg_{\mu\nu,\pi})_{,\pi}$). From (G9) og

$$\delta g_{\mu\nu} - \tau w^\pi g_{\mu\nu,\pi} = -\tau(w^\pi_{,\mu} g_{\pi\nu} + w^\pi_{,\nu} g_{\pi\mu} + w^\pi g_{\mu\nu,\pi}) + O(\tau^2) \tag{G23}$$

and (G21) becomes

$$\int_{y(\Omega)} \Delta\Theta^{\mu\nu}(w^\pi_{,\mu} g_{\pi\nu} + w^\pi_{,\nu} g_{\pi\mu} + w^\pi g_{\mu\nu,\pi})\, d^4y = O(\tau). \tag{G24}$$

We have dropped the carets from (G24), but must remember that it holds only on an extremal.

The left-hand side of (G24) is independent of τ. We can therefore take the limit $\tau \to 0$ and replace the right-hand side by zero. We write $\Delta\Theta^{\mu\nu} w^\pi_{,\mu} g_{\pi\nu} = (\Delta\Theta^{\mu\nu} w^\pi g_{\pi\nu})_{,\mu} - (\Delta\Theta^{\mu\nu} g_{\pi\nu})_{,\mu} w^\pi$, use Gauss's theorem and the symmetry of Θ, and find that

$$\int_{y(\Omega)} \Big[-2(\Delta\Theta^{\mu\nu} g_{\pi\nu})_{,\mu} + \Delta\Theta^{\mu\nu} g_{\mu\nu,\pi}\Big] w^\pi\, d^4y = 0. \tag{G25}$$

Since Ω can be chosen arbitrarily and the integrand of (G25) is a continuous function, it follows that the integrand is zero. Since the w^π also are arbitrary, og

$$-2(\Delta\Theta^{\mu\nu} g_{\pi\nu})_{,\mu} + \Delta\Theta^{\mu\nu} g_{\mu\nu,\pi} = 0. \tag{G26}$$

Eqs.(D9) and (D19) now imply that

$$\Theta^{\mu\nu}{}_{;\nu} = 0, \tag{G27}$$

where the semicolon denotes the covariant derivative corresponding to the metric g.

In special relativity, the components of the spacetime metric are absolute variables, and (G27) holds provided that there are no other absolute variables. In an inertial chart, the Christoffel symbols vanish, and the covariant divergence reduces to an ordinary divergence:

$$\Theta^{\mu\nu}{}_{,\nu} = 0. \tag{G28}$$

Similarly, in the vector theories of Chapter IX the components $\bar{g}_{\mu\nu}$ of the flat metric are the only absolute variables, and (G28) holds in a chart where $\bar{g}_{\mu\nu} = \eta_{\mu\nu}$. (In (G22) we must replace g by $\bar{g}$ and Δ by $\bar{\Delta}$; eq.(G27) is written as $\Theta^{\mu\nu}{}_{|\nu} = 0$, where the vertical bar denotes the covariant derivative corresponding to $\bar{g}$.)

There is another special case of our results which does *not* require that the $g_{\mu\nu}$ be absolute variables. As in Chapter IX, Section 3, we assume that the Lagrangian density can be written in the form $\mathcal{L} = \mathcal{L}_F + \mathcal{L}_G$, where $\mathcal{L}_F$ is a function of the dynamical variables $u_1, \ldots, u_P$ and their first partial derivatives, the $g_{\mu\nu}$ with $\mu \leq \nu$, and the $g_{\mu\nu,\pi}$ with $\mu \leq \nu$, and of no other variables. The $g_{\mu\nu}$ and $\mathcal{L}_G$ may be functions of the remaining variables $u_{P+1}, \ldots, u_{M+M'}$ and their partial derivatives, but *not* of $u_1, \ldots, u_P$ and their partial derivatives. The functions $\Delta^{-1}\mathcal{L}_F$ and $\Delta^{-1}\mathcal{L}_G$ are assumed to be separately invariant. (Usually, $\mathcal{L}_F$ is the covariant form of the special-relativistic Lagrangian density of the fields, while $\mathcal{L}_G$ is an additional term involving the gravitational fields.)

The argument that led to (G20) is unchanged if $\mathcal{L}$ is replaced by $\mathcal{L}_F$ (it requires only that $\Delta^{-1}\mathcal{L}_F$ be invariant). On an extremal, (G7) holds for the dynamical variables $u_A, A \in \{1, \ldots, P\}$. We identify $u_{P+1}, \ldots, u_{P+10}$ with the $g_{\mu\nu}$ for $\mu \leq \nu$, and instead of (G21) we find that on an extremal

$$\int_{y(\Omega)} \left[\mathcal{L}_{F g_{\mu\nu}} - (\mathcal{L}_{F g_{\mu\nu,\pi}})_{,\pi}\right](\delta g_{\mu\nu} - g_{\mu\nu,\pi} w^\pi \tau)\, d^4y = O(\tau^2), \tag{G29}$$

where we have again dropped the carets. We define a stress-momentum T by (cf. (G22))

$$\zeta \Delta T^{\mu\nu} = \hat{\mathcal{L}}_{F g_{\mu\nu}} - (\hat{\mathcal{L}}_{F g_{\mu\nu,\pi}})_{,\pi}, \tag{G30}$$

and prove by the same argument as for (G27) that

$$T^{\mu\nu}{}_{;\nu} = 0. \tag{G31}$$

The stress-momentum T is sometimes called the Landau and Lifshitz stress-momentum, although it was discovered by L. Rosenfeld [Rosenfeld 40].

References

This list is not exhaustive. It consists of books for further study, papers whose material is inadequately treated in books, and a few newer papers that the reader can use in pursuit of the latest developments.

Ashby, N. (ed.) (1989) *General relativity and gravitation* (Cambridge University Press, Cambridge).

Barker, B.M., Byrd, G.G., and O'Connell, R.F. (1986) Astrophys. J. **305**, 623.

Bertotti, B., Brill, D.R., and Krotkov, R. (1962) in *Gravitation: an introduction to current research,* ed. L. Witten (Wiley, New York).

Blanchet, L. and Damour, T. (1988) Phys. Rev. D**37**, 1610.

Brumberg, V.A. (1972) *Relativistic celestial mechanics* (Nauka, Moscow) (in Russian).

Brumberg, V.A. and Kopejkin, S.M. (1989) Nuovo Cimento B**103**, 63.

Caporali, A. (1981) Nuovo Cimento B**61**, 181, 205, and 213.

Chandrasekhar, S. and Contopoulos, G. (1967) Proc. Roy. Soc. (London) A**298**, 123.

Chandrasekhar, S. and Nutku, Y. (1969) Astrophys. J. **158**, 55.

Damour, T. and Deruelle, N. (1985) Ann. Inst. Henri Poincaré **43**, 107.

Damour, T. and Deruelle, N. (1986) Ann. Inst. Henri Poincaré **44**, 263.

Damour, T. and Schäfer, G. (1988) Nuovo Cimento B**101**, 127.

Eisenhart, L.P. (1926) *Riemannian geometry* (Princeton University Press, Princeton).

Gelfand, I.M. and Shilov, G.E. (1964) *Generalized functions,* Vol.1 (Academic Press, New York).

Hawking, S.W. and Israel, W. (eds.) (1987) *300 years of gravitation* (Cambridge University Press, Cambridge).

Larkin, P. (1974) *High Windows* (Faber and Faber, London).

Lighthill, M.J. (1958) *An introduction to Fourier analysis and generalised functions* (Cambridge University Press, Cambridge).

Lorentz, H.A. (1937) *The collected papers of H.A.Lorentz, vol* 5 (Nijhoff, The Hague).

Luther, G.G. and Towler, W.R. (1982) Phys. Rev. Letters **48** (121).

Misner, C.W., Thorne, K.S., and Wheeler, J.A. (1973) *Gravitation* (Freeman, San Francisco).

Nanda, S. (1976) Math. Proc. Camb. Phil. Soc. **79**, 533.

Newton, I. (1686) *Philosophiae naturalis principia mathematica* (London).

Poincaré, H. (1952) *Science and hypothesis* (Dover, New York).

Pound, R.V. and Snider, J.L. (1965) Phys. Rev. **140**, B788.

Rastall, P. (1960) Can. J. Phys. **38**, 975; (1979) Can. J. Phys. **57**, 944.

Rastall, P. (1979a) Phys. Rev. D**6**, 3357.

Rosenfeld, L. (1940) Mém. de l'acad. roy. de Belgique **18**, No.6.

Schäfer, G. (1985) Ann. Phys. (N.Y.) **161**, 81.

Schutz, B.F. (1985) *A first course in general relativity* (Cambridge University Press, Cambridge).

Smalley, L.L. (1983) J. Physics (London) A**16**, 2179.

Soffel, M.H. (1989) *Relativity in astrometry, celestial mechanics and geodesy* (Springer, Berlin).

Synge, J.L. (1958) *Relativity: the special theory* (North-Holland, Amsterdam).

Taff, L.G. (1985) *Celestial mechanics – a computational guide for the practitioner* (Wiley, New York).

Vessot, R.F.C. *et al.* (1980) Phys. Rev. Letters **45**, 2081.

Wald, R.M. (1984) *General Relativity* (University of Chicago Press, Chicago).

Weinberg, S. (1972) *Gravitation and cosmology* (Wiley, New York).

Westfall, R.S. (1971) *Force in Newtonian physics* (Macdonald, London and American Elsevier, New York).

Whittaker, E. (1951) *History of the theories of aether and electricity: the classical theories* (Nelson, London).

Will, C.M. (1981) *Theory and experiment in gravitational physics* (Cambridge University Press, Cambridge).

Index